Johanna Blomqvist, Ph.D.

THE MYSTERY OF WATER

A PHYSICIST'S DIVE INTO OUR DEEPEST ESSENCE

The Mystery of Water - A Physicist's Dive into Our Deepest Essence

ISBN (softcover): 978-952-69724-4-2
ISBN (hardcover): 978-952-69724-5-9
ISBN (EPUB): 978-952-69724-6-6
ISBN (PDF) 978-952-69724-8-0
ISBN (MP3): 978-952-69724-7-3

First paperback edition October 2024

Cover image created by an AI-based tool
Johanna's photo: Julia Alakulju

All images of book Johanna Blomqvist,
except image 6: Pixabay and image 20: Wikimedia Commons

Published by *Mindstream Publishing*
Helsinki, Finland, 2024
www.johannablomqvist.com

CONTENTS

As profound a reading experience as I can recollect. In *The Mystery of Water*, Johanna Blomqvist combines personal experiences like bathing in ice water, descriptions of diverse water-related spiritual phenomena like collective healing, and comprehensive accounts of numerous experiments showing that water is way beyond just H_2O; it's profoundly influenced by consciousness. Blomqvist offers perhaps the most comprehensive discussion I can recall of endlessly fascinating water-related phenomena, captivating the reader with descriptions of the vast array of seemingly anomalous but well documented behaviors ranging from water memory to holy water. I had trouble putting it down.

– Gerald H. Pollack, Ph.D., Professor, University of Washington; Editor-in-Chief, WATER

Johanna Blomqvist's *The Mystery of Water* is a refreshing look at this fundamental aspect of life and the planet. Full of intrigue and curiosity, Blomqvist takes the reader on an investigative journey that is sure to delight and surprise even the most educated reader. *The Mystery of Water* is a brilliant and innovative book that contributes to stretching our understanding of reality and ourselves. I highly recommend this book!

– Natalie Dyer, Ph.D. in Neuroscience, Research Scientist

THE MYSTERY OF WATER?

What do we really know about water?

- About 70% of the total mass of our bodies is water. 99% of all the molecules in our bodies are water molecules.
- Jellyfish consist almost entirely of water and have no heart or brain. How is it possible for them to still be living and responsive beings?

Why does water behave differently from other liquids?

- Usually, the solid state of a substance is heavier than liquid and sinks into the liquid. Ice, however, is lighter than liquid water and floats on the surface of water.

Why is water liquid at the Earth's average temperature of about 59 °F (15 °C)? Other similar compounds are gases.

- In fact, water differs completely from other liquids: it is known to have more than 70 anomalous properties.

Solving the mystery of water could also help us understand ourselves.

FOREWORD AT THE WELLSPRING

"Water is to me, I confess,
a phenomenon which continually
awakens new feelings of wonder
as often as I view it."

– Physicist Michael Faraday (1791–1867)

In many spiritual traditions, especially the older ones, it is customary for a teacher, a master, to guide a student on the path of spiritual growth to encounter and realize essential teachings. It is said that the teacher appears when the student is ready. My own teacher appeared unexpectedly and in a surprising form.

Only one month before the due date of my book *Hyperreality*, I left for a cruise to Stockholm, Sweden with my family. The book was almost ready, but of course it needed lots of editing. I thought, however, that a few days' rest before the last push might be a good idea.

The trip to Stockholm was rather peaceful but the return home was wilder than I have ever experienced, even though I have been on many cruises during my life. When the ship left Stockholm, the captain announced that the waves may reach the height of 13–23 feet (4–7 meters). I checked later from the wind maps: the wind had been from 65–80 feet per second (20 to 25 m/s) which in Finland is classified as high storm wind. Our cabin was in the bow of the ship and the rocking there was naturally strongest.

I've never been seasick on a ship before due to the wind and waves, but now I couldn't bear them. Perhaps reading also increased my nausea. For a while, I read Heikki Peltola's book *Whole and One*, since I had promised to give my comments to Heikki in a few days. At 2 a.m. I was so sick that I couldn't hold it in anymore, and I rushed to the toilet and threw up. The next two hours I spent there on the floor of the cabin's toilet. Those hours were indeed long. I still remember the white rectangular tiles of the toilet's walls, the shower curtain that had already seen its best days and the white and thick bath towel which I used as my blanket since it felt disgusting to bring my blanket to the toilet floor. And I was so tired, too. I was really unwell, and when you are tired in the middle of the night everything seems much worse. It was not only the rocking inside my head and in my stomach, it felt as if my whole body was moving slower than my own spirit. My body seemed very heavy.

At times images flashed in my mind: events and feelings linked to those events. I wondered what this episode was about. It seemed to me it was somehow about control, the mind and my fears. It felt as if something had been activated in me, and now something from the past was being released. I pondered what I feared and what I was figuratively, and quite literally, expelling from myself. The thought of not being able to manage, to keep everything together seemed particularly difficult. I felt this was related to some other moment, probably in the past. On the other hand, it seemed to also connect to this very moment, when I was struggling to finish my book.

All my senses seemed to be activated, sensitized. I could hear the sounds of the wind getting louder. The more tired and closer to the borderlands of sleep I was, the more images flashed in my eyes about storms, floods, Estonia and Titanic, Atlantis, all those upheavals and tsunamis that have destroyed villages, cities and even massive megalithic structures in different parts of the world. This storm was unlikely to be the end of me, but it felt like a portal to another moment somewhere. I was a little amused though, because after all, I was merely traveling on a ship from Sweden. For us Finns these kinds of short cruises from Helsinki to Stockholm are quite common, and especially a decade or two ago people made this voyage many times a year.

I felt like I was having a conversation with water. It seemed to know everything and have all the answers. It kept changing; first, appearing stormy in my mind, the next moment gentle and calm. As if it would have shown me different sides of itself, but also of myself. Feeling like I was traveling with it through time, experiencing everything it had experienced, I finally fell asleep around 5 a.m. By then I had finally let go of everything – all the fears, thoughts, and bodily sensations and, in my mind, surrendered to the waves.

Morning dawned and the storm abated. The next day, I felt tired and grateful. The feeling of misery during the night had changed into gratitude for the experience and daring to face my fears and emotions. I also realized that one meaningful part of this was water. I had made a new kind of connection with it and I was sure that our journey together would continue. Later, I also understood that water is now my teacher, who also guides me where to focus my attention.

Over the next few months and the next couple of years, the world was turned upside down, and water became increasingly important to me. The more I got to know it, the more I also realized how little we know about water and in how many ways it is a mystery. And if water is a mystery, how could we even understand ourselves, because we are mostly water!

I wrote in my book, *Hyperreality*, about the attempt to understand reality: *"The Being, understanding of which I am seeking, is myself. The essential point is who is asking."* Maybe water gives us answers? Maybe it leads us toward the fountain of everything, where all knowledge and wisdom arises?

Two storylines flow in this book side by side. One tries to understand what water is empirically or objectively, and the other to understand it more experientially, that is, subjectively. The goal of this book is to actually find answers so as to understand ourselves better. So, listen to your thoughts and, from time to time, your emotions while reading this book. Let them carry you, because you and water are one. Water can be seen as a metaphor or as a model for ourselves – what works for water probably also works for us, too. This book is about us.

I address water from many perspectives, and of course, as a physicist, I also approach the subject from the perspective of the natural sciences. You don't need to be familiar with science or physics and chemistry to understand this book. If you know what a molecule is and that water has different states, you are already a long way toward understanding. There is a glossary at the end of the book, where I explain some terms and concepts in more detail. The same is also true for the references, which are marked only when necessary. At the end of the book, you will find a reference list where you can read more on the topics.

There are many mysteries associated with water. One can only wonder how it is possible that a compound so close to us is associated with so much that is not understood or is even ignored. In the book, we walk on the edge of science from a deep effort to truly understand and find answers to the questions I posed at the beginning of the book. Read the book with an open mind, with the attitude of a curious researcher, without dismissing what you read. Be ready to shake up your own thinking and possibly also to question the prevailing dogma. We usually prefer to stick to the view we have learned and grown into during our lives, and we don't often question it unless we have to. It may be that we are afraid that our whole life will be ruined: *"If this doesn't work like this, then maybe this won't work either."*

I will tell you in this book about different kinds of approaches to water, different kinds of research carried out by scholars who are devoted to studying water from the deep wish to understand. I will also tell you about water-related stories and experiences. You will notice that some themes are repeated in different contexts. Maybe we could learn something new from those? By drawing together all that is known about water I wish to be able to help in forming a complete view about this vitally important compound and element.

I argue that to understand ourselves better we should try to understand water better. Not only because we are mostly water, but also because even though the properties of water differ so much from those of other compounds, nature always tries to organize itself to its simplest form. Water should be similarly understandable. There is nothing useless in nature. Nature also repeats itself. Understanding water may help us forward in our spiritual journey.

Before we familiarize ourselves more with the mystery of water, we start with the myths of water, according to which water has always existed. Next, we will study what water really is, both chemically and physically. I will address water also experientially because its essence can't be reached in other ways. I come from Finland, a Land of Thousands of Lakes, and I must say that we are "water people." The essential place of water in our lives is self-evident; we live through water. Many of us, like myself, live by the sea, a river or lake, and when it is summertime, we spend our days doing different kinds of water activities. However, it also seems that we have forgotten how important water really is to us. With this book, I hope I can bring the importance of water to our awareness, so that we will appreciate it and take care of it more.

Understanding water may turn out to be significant for both science and our well-being. I will tell you how water behaves at interfaces, whether water can remember and how we affect water. On this journey, I will also address holy water and carry out my own water experiments. During my writing process, I let the water guide my journey. Diving deeper into the water has given me an understanding of reality on a completely different level and depth than one could imagine. You can read the chapters of the book in order or let the water carry you in the same way as when I was writing the book. I felt the water lead me forward. You have to learn to understand water on many different levels. This understanding can also be increased by letting the text lead you, just like water flows, transforms and sometimes takes you in surprising directions to see and experience something new.

Welcome to a research journey into the mystery of water – our deepest essence! Let's travel together to the source of everything – to the wellspring! I hope you enjoy the book!

July 2024,

Johanna Blomqvist

CHAPTER 1. WATER IS THE BEGINNING AND END

"All things are from water and all things are resolved into water."

– Thales of Miletus,

a philosopher and astronomer,

circa 600 BC

~

There is a story that tells how Thales of Miletus was once looking at the sky while walking and, when counting the stars, he fell into a well. While falling into the water he realized the importance of water: water is the beginning of everything. The Greek word *arche*, which means 'beginning' also means 'source from and in which everything happens' – the strongest and all-encompassing which reveals all. Aristotle used *arche* to mean first principle, or basic principle of knowledge from which others in a science are derived. With water everything receives existence and becomes visible. Thales of Miletus saw that water also included time and experience, and the passage of time was measured with water clocks, in Greek *clepsydra*. Water and time also contained wisdom.

Water is the beginning and it is also the end. Water is both loved and feared. We need it, but too much of it is too much. It is present in our life all the time, and it has always been. Many often take it for granted because we've had it in abundance. Water is probably the oldest cultural symbol, and it is mentioned in most myths about the creation of the world and in natural philosophies dominant before the development of science.

WATER IN THE BEGINNING

In Judeo-Christian mythology, the creation of the world is described in Genesis. The first book of Moses (1:1–2) says: *"In the beginning God created the heavens and the earth. Now the earth was formless and void, and darkness was over the surface of the deep. And the Spirit of God was hovering over the surface of the waters."*

The earth was thus created, but water existed already before everything else. After the creation of the world, God's first act was to bless the waters. The Catechism states that *"since the beginning of the world, water, so humble and wonderful a creature, has been the source of life and fruitfulness."* And just as the world has been created and born into water, we also grow in the water of the womb, says the Bible.

Water plays an essential part in the folklore of many different cultures. And just as in Genesis, water has not been created but has always existed. In the Kalevala, also referred to as Karelian mythology, the national epic of my home country Finland, the world is being created from the fragments of an egg laid by a scaup (lat. *Aythua marila*) on the knee of Ilmatar, goddess of the air. When Ilmatar moves her knee, six golden eggs and one iron egg laid by the scaup roll into the primordial ocean and break into pieces. These pieces form the land, sky, stars, sun, moon and clouds. The goddess of the waters, *Veden emonen*, creates capes, bays, beaches, shallows and deep places in the sea. Väinämöinen, too, the hero and main character of Kalevala, is born from the goddess of the water and drifts on the waves until the wind carries him to the shore. *Väinä*, the beginning of the name Väinämöinen, means wide and slowly flowing river or strait of the ocean.

The realm of the waters in Kalevala mythology is an entirely separate world of its own, ruled by *Ahti* and *Vellamo*. It was customary to ask for good catches of fish from Ahti and, in return, one promised to behave well while afloat, without making noise. Vellamo was also asked for good luck in fishing. Vellamo ruled the waves and storms, giving favorable winds to seamen. There was also a wide range of other water people: water maidens, nymphs, spring elfs and other spirits of the waters, like *Vetehinen*. There were also evil water spirits, such as Hiisi, and a sea monster called *Iku-Turso*.

Myths about an egg as the beginning of everything can be found in other cultures as well. Water, or *ap* in Sanskrit, is according to the Indian Vedic texts as old as the universe, which also was born from an egg. The world is said originally to be "water without light," which is the same as in Genesis. Water, since it is so primal, is according to Vedic writings divine. It was said to bring peace, happiness, abundance, longevity and health.

According to Indian Upanishad writings there was Nothing in the beginning. This Nothing however wished for selfness and created for itself a mind, the movements of which formed the water and, from the foam, the land. When Nothing took a physical form, a human was born. The human was lonely and divided itself into two pieces, forming a man and wife, out of which all the people in the world originate.

Similarly, as in the Upanishads, in Ancient Greece it was thought that there had been something permanent and changeless that was the beginning of everything. Plato describes in *Timaeus* how the world of appearance was created based on the eternal, unchanging true reality, the world of Forms. Since nothing is born without a reason, Plato introduced the idea of a demiurge, the creator of everything, god, behind 'the everything'. The changeless model of the world, the world of Forms, is in his mind. Matter is also born into the world, lacking the stability of **what is** changeless, since all form elements – earth, air, fire and water – are in constant motion, mixed with each other and the formless.

According to North American Cherokee folklore, the world was first simply water and darkness, thus exactly the same as in Genesis and the Vedic texts. All the animals lived in the vault of the sky in the world called *Galúnlati*. When it was overpopulated, animals had to move down to the world. They found mud from the bottom of the ocean and that mud swelled until it became the island called Earth. This island is attached from its four corners to ropes hanging from the sky. These ropes will some day break, and the earth will drop back to the ocean.

In the creation myths of the Popol Vuh texts of the Mesoamerican Mayas, the universe was first divided into the ageless ocean and infinite sky. Water divided the universe into the Upper World (the realm of gods and spirits) and the Lower World (the realm of power animals, ancestral spirits, gods of the dead people). The path between these two always went through water. The World was born when the earth rose from the great ocean, and the plants and heavenly bodies were born at the same time. There were many unsuccessful trials while creating a man. First, the animals were created, but since they were unable to speak or sacrifice to the gods, one tried to create man from the clay, which however dissolved to the rain. Man made of wood wasn't able to speak, and these creatures were formed into monkeys. Last, the realm of water was asked for help and a man was born of water. Death was called "going to the water." The surface of the Lower World was water, and one could reach it with canoes or fountains and rivers flowing in the caves. Gods paddling in their canoes carried the souls of the dying to Xibalba, which existed below the water.

Ancient Mesopotamia had several creation myths since there were many different cultures in the area. However, water and the great flood link them all together. In the creation myths around the world water is fundamental. It was of such importance that in many cultures there are several gods of water. In Egypt, for example, there were about a dozen gods of water. The Nile and its flooding had several gods: *Anuket, Hapi, Khnum, Satet* and *Sobek*. Death and water were also tightly bound to each other. *Nephthys*, for example, was the goddess of rivers, death, mourning and night, and *Osiris* was first the god of water and vegetation, but later god of death and the afterlife. Besides these, water, humidity and fertility had their own goddess called *Tefnut* and primordial waters goddess *Nu* or *Nun*. Nu was also the goddess of the sky, stars, cosmos, mothers, astronomy and the whole universe. She was often depicted carrying water in pottery above her head. Nu was considered the first of all the gods and also the creator and destroyer of reality. Water and space were seen to have a connection. Both were equally deep, eternal, unattainable and at the same time fascinating and inviting.

One exception must be mentioned, though, and that is China. While in the myths of other cultures water was seen as fundamental in the creation of the world, in Chinese folktales there was *qi* or life force in the beginning of everything. In these myths, *qi* was thought to rule matter, time and space. At the time of creation, dualism – masculinity and femininity, Yin and Yang and all the other binary forms – were born from oneness.

CATACLYSM MYTHS

Water was also the end. Many cataclysm myths describe great catastrophes – floods – that destroyed massive constructions, whole ancient cities or civilizations. Myths about cataclysm seem to exist in all cultures and in every continent. The world's oldest epic, Gilgamesh, from around 2000–1600 BC, describes a water cataclysm. The epic tells about the heroic acts of King Gilgameš and his search for eternal life in ancient Mesopotamia, situated in the area of current-day Iraq, in the valley of the Euphrates and Tigris rivers. The cataclysm story of Gilgameš describes a week long storm, which made waters rise and darkened the sky and land. Utnapištim, the hero of Gilgamesh had changed his house before the storm to a giant ark and stored all the living seeds as he was guided. After seven days the highest peak of the mountain finally emerged – Utnapištim sent first a dove, then a swallow to search for the land. When both returned, he also sent a raven. The raven didn't return and the land was found – Utnapištim was saved together with his wife. The scaup flying on the top of the endless water in Kalevala is said to refer to a very similar kind of great flood.

The story of Gilgameš is also very similar to the story of Noah's ark in the Bible, where a rainstorm covered the earth for 40 days. In Greek mythology the ark was built by Deukalion, who together with his wife are the only survivors from a flood lasting nine days.

Atlantis and its disappearance into the depths of the sea is perhaps the most precisely timed cataclysm myth. Atlantis is mentioned for the first time in the writings of the Greek philosopher Plato. Plato describes the war between those "living beyond the 'Pillars of Hercules' and the Athenians." The Pillars of Hercules is thought to mean the Straits of Gibraltar, sometimes the areas outside Athens in general. Non-Athenians were led by the kings of the island of Atlantis. Plato tells about the kingdom of Atlantis and its destruction in *Timaeus*: *"In this island of Atlantis there was a great and wonderful empire which had rule over the whole island and several others, and over parts of the continent […] But afterwards there occurred violent earthquakes and floods; and in a single day and night of misfortune all your warlike men in a body sank into the earth, and the island of Atlantis in like manner disappeared in the depths of the sea. For which reason the sea in those parts is impassable and impenetrable, because there is a shoal of mud in the way; and this was caused by the subsidence of the island."*

Atlantis sank around 9600 BC during the great flood in an earthquake, which seemed to cause a mud barrier preventing sailing in those areas.

Atlantis hasn't been found, even though there have even been conferences of scholars seeking to find it. Several locations have been suggested as potential sites of Atlantis. In autumn 2018, Jimmy Corsetti on the Bright Insight YouTube channel presented an interesting suggestion: "The eye of the

Sahara," also known as the Richat Structure would perfectly fit the descriptions of Atlantis. The area is about 25 miles (40 kilometers) wide and resembles an eye in the Sahara Desert, Mauritania. It greatly resembles descriptions of Atlantis, containing alternately mountains and low areas like water canals. It also looks like these areas have been underneath seawater, as fossils of sea creatures have been found there. According to the prevailing understanding, it is considered that the birth of the eye of Sahara is due to the breakup of the supercontinent Pangaea, when the waters of the Atlantic Ocean flooded to the area. No man-made structures or anything that would indicate it was once inhabited have been found. Maybe the real Atlantis is somewhere else – in the bottom of the Atlantic Ocean or fragmented around the Atlantic and its coasts.

In any case, the year 9600 BC given by Plato is considered erroneous because, according to the prevailing understanding, there have not been any high cultures this old, and in Northern Africa people were living in hunter-gatherer societies. But why would Plato have made such an error? Other events and locations Plato described have been accurate, and they have been verified as correct through other routes.

In actuality, more and more facts seem to lead to the conclusion that the year Plato proposed might well be true. It seems that about 11500 years ago, which is around that year, there was a great catastrophe on earth which caused the seawater levels to suddenly rise. Large areas were flooded and also destroyed. According to some studies these events also included high temperatures, and many buildings and statues were burnt black. Maybe it was a massive sunburst? There is evidence of water rising in Egypt to support this conclusion. According to several Egyptologists one can see marks of water erosion in the sides of the Sphinx. These marks would have been caused by great floods which occurred around those years. If this is true, it would mean that the Sphinx would be older than has been assumed. Current belief has been that the Sphinx was not built until the reign of the pharaoh Khafre, in 2558–2532 BC.

Our understanding of when the earliest high cultures evolved seems to be changing. Only a few years ago, Göbekli Tepe, situated in Turkey (Türkiye), was found. It was built around 10,800 – 9500 BC. There are massive megalithic constructions in the area including gigantic letter T-shaped stones with animal figures. Megalithic constructions like that are known to exist in other places as well, and they seem to be even older than Göbekli Tepe.

Gunung Padang, for example, is an ancient megalithic temple situated in Indonesia currently covered with massive hexagonal stone columns collapsed everywhere around the area. The temple is located at 885 meters above sea level up on the hill, currently deep in a jungle. In the excavations, layer after layer of passages, cavities and manmade structures have been found inside the hill. With carbon dating, the age of the place has been approximated to be from 22,000 BC, making it older than any other known man-made building. What is even more fascinating is that a fountain has been found in the heart of the temple. Its water was probably used in different ceremonies. Fountains have been found at the center of many other temples, too, as if they are the heart of the place. Similar techniques of edging slopes with stones used in Gunung Padang were used in Macchu Picchu in Peru, among others. In the Macchu Picchu area, the oldest structures have been finished most accurately: the seams of huge megalithic stone built walls are still tight. Single stones which can hardly be lifted even with today's technology, in those places up in the mountain walls are still extremely close to each other. The newer stone buildings constructed by the Inca can easily be recognized, and in many of these, megalithic stones have been reused as building material.

Of course, I am not an archaeologist or geologist, and my intention here is only to awaken your curiosity to consider different possibilities. Most likely, we don't know yet the whole truth and must do more research. Maybe the myths about cataclysms are not only myths? What if one of the properties of our world is cyclicality? Like waves of the ocean, the water level rises in regular intervals, wiping away old structures. Maybe this is the reason why the ancient peoples built underground caves and networks of corridors, places to survive longer? Could it be that with their stone wall carvings about ancient destructive events and floods they wanted to tell future generations about their destiny so that we would be able to prepare ourselves and understand that similar tidal waves would probably come again? There seems to be a connection too between the creation and cataclysm myths. Maybe the creation myths tell how the world was reborn after the cataclysm?

Water can be at the same time the end of the world and also the beginning. It can be destructive but also protective, carrying something from the old world to the new.

Cataclysm myths can also be interpreted as a kind of a collective phenomenon emerging from our unconscious, as Carl Jung describes in his theory of archetypes. The cataclysm is often preceded in stories by war, injustice and corruption. As it tumbles, water cleanses and resets civilization back to common constitutive and primal values and principles. In *Hyperreality*, I presented five principles – Consciousness, Oneness, Interaction, Freedom and Complementarity – that can be seen as life-guiding instructions based on modern physics. Through modern physics it is possible to end up at the following conclusion about the eternal and primal nature of consciousness: we are not separate but instead all of the same oneness and in connection with each other all the time. There are no opposites because everything leads to one. Reality can be created from the same oneness and from the same fundamental values such as love, goodness and justice.

~

A few years ago, I studied hypnosis and graduated as a hypnosis practitioner. I was interested in understanding the human mind and consciousness. I also familiarized myself with the so-called QHHT (Quantum Healing Hypnosis Technique), which is a method developed by hypnotherapist Dolores Cannon. She used it for almost five decades and wrote about her experiences in 17 books. QHHT is a technique to deepen the hypnosis and connection with the inner self, which may also be called higher consciousness or the subconscious. The hypnosis practitioner works as a guide in this journey, helping the person to relax and notice the essential issues. During this journey you can end up seeing another moment in your life, a different form, or a past life, whatever helps you in the best possible way to notice the essential, or what is needed in your life right now. With QHHT you can look for the answers to questions you have in your mind, helping you to heal and grow.

The following story relates my experience of the QHHT hypnosis journey. The hypnosis was stored as an audio recording and the text has been written based on that. My guide was Marjut Saariniemi, a QHHT practitioner that has been conducting these sessions for years. The hypnosis was done years ago, long before I had any hunch that I would be writing a book about water. Maybe my hypnosis journey tells about a cataclysm as well?

Cataclysm hypnosis journey

I first felt my feet, and how they didn't move. I tried to free them, but I was knee-deep frozen in cold, watery clay, sinking. I felt a conflict: somewhere else, in another situation, the soft clay on my skin would have felt good. But not now, because the clay pulled me deeper and deeper. I felt the ground disappearing beneath me.

My senses were activated. The feeling is extraordinary, like you've been suddenly dropped into a movie or a video game, but your consciousness would have been too slow to understand that instantly. Even though my body was relaxed, the first feeling was as if I were panicking, and my senses arrived later, as if they had been at the start of a movie and one by one were activated to become this character. I noticed that I didn't hear anything, but then I realized that I didn't hear anything because the noise of the water overwhelmed everything else. I heard its noise, but I didn't see anything – at least at first. After a bit I realized that it was actually so dark that I could only see occasionally. The wind was also blowing hard, and the sounds of palm trees being whipped by the wind had never sounded so loud before. A large body of water was approaching. I grabbed a tree's roots that had been exposed from under the clay. In my mind a fear had come for my loved ones, mom, brother, the whole village... I hoped they all had made it out of the flood unharmed. I felt a sense of guilt, because I had run to the forest in the evening without permission. Could I have helped them, if I had been home? That's what I thought, because I was the oldest child. Everything poured into my mind as one, my whole life and all its stages. Interestingly, at the same time I knew that I was watching someone else's life other than my own. I couldn't focus on this thought longer, because the next moment a huge wave caught me and I ended up getting swept away with it. It swept with such a force that it was hard to tell if I was different from it or the same. I couldn't think of saving the others at that moment, because I was the one in need of saving. Even if I was a good swimmer, a feeling came that this wouldn't end well.

I swayed back, forth and everywhere with the body of water. I could see almost nothing, because the sky was pitch black, and at times mud came into my eyes, ears and mouth. I didn't know which way to head. I just heard the water's noise and sometimes it hit on my body, when I was bumped by trees, rocks and probably even some buildings' bits, at least that's what I think they were. I couldn't fight back, that would only have exhausted me. The water was the driver and I was only going with its journey. In the background I experienced a fear, about where I'd end up and what would happen next. I already felt death.

The journey ended surprisingly. I hit my head. I was somehow stuck to a smooth wall, but the water was still pushing me. The wall was probably a dam. I couldn't move or breathe, and I felt how my energy was fading. Fighting against the water felt impossible. How is it possible that my most loved element, water, now acted against how I normally felt it?

As my energy was fading, I suddenly got new superpowers. I noticed that I could speak with the water. Its rumbles and noises faded, and I just heard a whisper that filled the whole space, that couldn't be avoided. It said: *"Don't fight, because that's useless. When you let go, you'll hear my voice better. That's how it's supposed to be."*

Letting go wasn't hard, because my energy faded fast. I was strong, but not strong enough. I had always tried to control my world, the village that I lived in, our small house and kids, my siblings, and just when I had learnt to control, I had to let go. I felt that with control I also lost all my roles. There weren't any options, I didn't have the strength to fight back anymore. I let go.

I felt my life flashing before my eyes. Every moment of my short life appeared in front of me bright. I was shown moments of my life that weren't always so brilliant. I had lived in a village surrounded by a jungle, where we were happy, even if we had almost nothing. We walked without shoes and often didn't have food. My father had died. As the oldest child I decided things for my mother and siblings, and I also controlled their lives. They weren't always happy with my decisions. Now I saw and felt things from my own but also from others' perspectives, and it wasn't an easy experience. Had I caused this for others? I wish I had known sooner!

As my physical powers were fading, I felt stronger. The body that was fighting against the water or the difficulties of life wasn't real, but rather the person, the spirit, that was telling this story was. This awareness was something more permanent and primary, before anything else. That spirit is part of reality, perhaps something more real than what can be found, and that embodies something all encompassing and surpassing consciousness. I slid up and now watched my body down below. My small body had been squished between the now broken dam and fallen-down pillars and as the water pushed it was even more stuck. A huge body of water had come far from the oceans to our village and had destroyed the surrounding almost nonexistent rampart. Up above it was easier to see that our whole village was under the water's control too. There wouldn't have been any chance of surviving even there. Everywhere was just water. Now it felt remarkable to see how I had throughout my life tried to take care of myself and others. It was important to also see the deep connection with water. How important a role it played before, when we lived closer to nature. It has been a part of everything, and how every day and life were planned around water.

I floated away from the story, it stayed behind and I knew that I had witnessed the essential, everything that I needed to for this moment. I slid between space, infinity all around me, until suddenly the motion stopped and I fell on some type of round, flat surface. When I looked around, the corners of the surface around me had light figures. I tried to count the figures, but they were moving; there were either 9 or 11 of them. I saw the characters as tall light figures almost like they had cloaks on. I couldn't see them more clearly, but I got the feeling they were friendly. I could have a conversation with them, even though we weren't talking. Information seemed to be passing telepathically. I was with them getting answers, except that they told me that I didn't truly need their answers, because we are led through our own journeys in life. The reason I was with them was energy – I was getting charged there and being transported to a new level of energy. When I, of course, asked the meaning of life, I got the answer that we are here to learn and experience life as matter, as physical beings. We should learn to live life with the flow like water, in constant change and often in uncertainty – to live from our hearts.

~

Did I invent the story in my mind? Was it just a product of imagination, or have I really lived such a life before? Answering these questions is not essential – what is important is the effect that experiencing the story has on oneself. The experience felt very real, like a memory of my own. The hypnosis helped me understand my relationship with water and why my experiences with water have sometimes involved fear as well. The story also had a recurring theme of the need for control and how one must learn to let go of that too. The hypnosis helped me focus on what is essential. It felt like I reached that narrator's voice inside me, who has all the answers. The everyday reality with all its distractions and irrelevancies fades away, leaving only what is important. Returning from hypnosis felt like waking up from a dream back to everyday life.

~

Water seems to hold answers to many questions. Yet, it remains a mystery to us. Why is this so?

CHAPTER 2. WHY IS WATER A MYSTERY?

"Water is H_2O, hydrogen two parts,
oxygen one, but there is also a third thing
that makes it water and
nobody knows what it is."

– D. H. Lawrence, Author (1885–1930)

As strange as it sounds, water – a compound that constitutes over 70% of our bodies and covers nearly the same proportion of our planet – is the most common, yet the least understood chemical compound and liquid. No other compound behaves in such a peculiar and anomalously different manner as water does. It is known to have over 70 anomalous properties, including its states of matter, density, and material, as well as its thermodynamic and physical properties. The number of anomalous properties continues to grow as more is learned about water. Moreover, water's properties are so fundamental that if it did not behave this way, life as we know it would not exist!

Therefore, water is a mystery in various ways regarding its properties, but why hasn't its mystery been unraveled? Why is water still a mystery today?

Greatest anomalies of water

- As other liquids freeze into solids, their density increases, but not so with water. Ice is thus lighter than liquid water, which is why it floats on the surface of water. This is indeed needed for aquatic life!
- Water stores and transports heat very well. It has a particularly high heat capacity and thermal conductivity. This is why heat is retained well in

the body, and we can regulate body temperature. Water also stores heat as the sun warms the seas and lakes, and for example, the Gulf Stream transports heat from the equator to the north.

- Water has a much higher melting and boiling point than other similar compounds, such as the hydrides H_2S, H_2Se, and H_2Te. Other compounds are gases at Earth's average temperature of about 60–70 °F (15–20 °C) and at 1 atm pressure, while water is a liquid. If this were not the case, and water behaved like the other compounds, we would not exist.

DON'T WE KNOW EVERYTHING ABOUT WATER ALREADY?

Even many scientists have the belief that water, as such a common compound, is surely well understood and researched, and that there are no longer any scientific challenges associated with it. To some extent, then, water is a mystery because we do not even consider that there could still be something to uncover. However, water has many peculiar properties that are not understood. And some of these are not even deemed worth mentioning.

Research on water has long been divided. On the one hand, the structure, properties, and interactions of water are being studied and understood increasingly well. At the beginning of the millennium, I participated in a project that studied the water molecule through computer simulations. Computationally, it is possible to study, for instance, scenarios that are difficult or expensive to create experimentally.

There is also water research that is usually presented only at fringe science events or spiritual fairs – in other words, the same venues where topics might include UFOs, energy healing, past lives, and other subjects considered pseudoscientific or part of parapsychology. The idea that water might have some kind of memory is one example of such a topic. Water is understood to be different and deviating from other compounds, but often only to a certain extent. It is thought that water must fit into prevailing models and theories, but at the same time, it is forgotten that there are also other mysteries in modern science. Prevailing models of reality completely overlook consciousness, even though we know that reality is both physical and mental. The general assumption seems to be that what should be understood and known about water has already been discovered, that everything else is nonsense, and nothing new can be found out about water.

To understand how this situation has come about, one must go back in time a bit.

WATER SHOULD BE STUDIED DIFFERENTLY

At the beginning of the 20th century, quantum physics and the theory of relativity brought about tremendous revolutions in the natural sciences, which in turn have influenced the developments of the last hundred years. Technical applications of quantum physics, such as mobile phones, computers, and others, have conquered the world. I could not write this book on my laptop, from which it is saved to the cloud and later sent to my publisher and then to book distributions services, without the insights of quantum physics.

An even bigger revolution than what has occurred in applications has been the change in thinking brought about by modern physics. This change is actually still ongoing, as understanding the implications requires an entirely new way of thinking. The observer effect is one example of this: the observer, or the examiner of the experiment, always affects the outcome of the experiment, which is a truly revolutionary idea. It is impossible to obtain objective results from quantum physics experiments, as the observer – one might even say – creates the reality they observe. Consequently, the results had to be viewed in a completely new way due to quantum physics, taking into account the experimental setup as a whole and also the philosophy behind the results by asking what the results really mean. Early 20th-century scientists also understood that the whole is more important than its parts. Understanding water was seen as important, as water is everywhere. The thoughts of physicist and philosopher David Bohm are an example of such a viewpoint: each system under study is seen as a whole, and therefore initially, this whole is examined rather than its parts, in order to seek a deeper order, the so-called *implicate order*.

However, the ideas of materialism began to gain strength. Materialism emphasized reduction, that is, breaking down the subject of study into ever

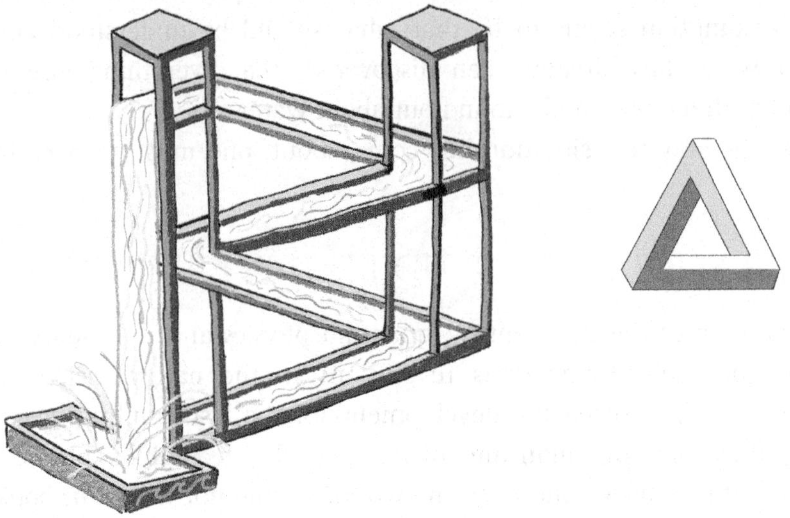

Figure 1. The impossible waterfall and on the right the Penrose triangle.

smaller parts. It was thought that as devices improve, increasingly precise information would be obtained, leading to a better understanding of the behavior of matter, which in turn would bring us closer to scientific truth. Holistic thinking had to give way when it was believed that answers could be found on a smaller scale. This way of thinking gradually expanded to other disciplines, focusing more on the individual parts and not holistically on the whole. Such reductionism is characteristic of the prevailing thinking in Western countries, scientific materialism, and it continues to live strongly. It is thought that answers are found on a smaller scale, or at least not by examining the whole more broadly, more holistically. This did not bode well for water research. The same reductive scrutiny cannot be extended to all phenomena.

There were also researchers who saw problems with this approach. Biochemist and Nobel Laureate in Medicine in 1937 for his discoveries in connection with the biological combustion processes, with special reference to vitamin C and the citric acid cycle, Albert Szent-Györgyi, believed that water is the foundation of all living things. According to him, biologists face challenges in understanding life and living systems because the tendency is to focus on the material world, which usually results in water being overlooked.

To solve the mystery of water, it should be examined in a different way from what we are currently accustomed to. Everything may not be as it appears at first glance. When you look closely at Figure 1, you will notice that the waterfall is impossible. The image is based on the waterfall lithographic work by Dutch artist M.C. Escher. A visual paradox arises from conflicting proportions. The aqueduct carrying water to the waterfall is structurally in the shape of two Penrose triangles. The triangle is named after the discovery by psychiatrist Lionel Penrose and his son, physicist Sir Roger Penrose, describing the triangle as *"impossibility in its purest form."*

Perhaps a different approach to water could reveal something new? Water is so exceptional, anomalous, that it most likely also needs to be approached differently.

WATER IS A HOT TOPIC (AND NOT IN A GOOD WAY)

The view of Gerald Pollack, a professor of bioengineering at the University of Washington and perhaps the most well-known water researcher of the 21st century, is that the study of water has gained a bad reputation over the last hundred years, especially due to two incidents that occurred between the 1960s and 1990s. Although some time has passed since these incidents, their impact persists. Now, however, due to the bad reputation, it is more difficult to obtain funding for water-related research. Many researchers also fear being stigmatized as they approach a subject whose investigators have been discredited one by one. Unfortunately, these cases are not unique – often a new, deviant, or unknown topic is initially dismissed because it does not fit into the prevailing paradigm. We will return to Pollack's water studies later in this book.

~

The polywater incident

Russian scientists Nikolai Fedyakin and Boris Derjaguin made an interesting discovery in Russia in the 1960s: water was heated and then cooled in thin glass capillary tubes, it behaved very differently from ordinary water. The molecules of water vibrated in a different manner, the density of the water was exceptionally high, and moreover, it was difficult to freeze or vaporize. The properties of water in capillary tubes became entirely exceptional. Because the properties also suggested a stability and permanence similar to polymers, the chemists considered it as some sort of polymer water and started to call it "polywater."

Naturally, the discovery received publicity and researchers were excited, as it was something revolutionary – perhaps even a new state of water. In fact, the idea of a new state of water was not new at the time, since chemist Sir William Hardy had suggested in the 1920s that water could have a fourth or even fifth state. In the case of Fedyakin and Derjaguin's discovery, it turned out that impurities in the water, such as salts and silicon released from the glass capillary tubes, were involved in the properties reported by the researchers. It is quite common for samples to contain impurities, and that is also why repeated measurements are made to eliminate various errors. It is particularly difficult to obtain a pure sample of water, as water is a natural solvent for almost everything, meaning it tends to absorb all sorts of foreign molecules into itself. Impurities are, in fact, a natural property of water. If they are present in water in limited amounts, it does not necessarily mean that the phenomenon should be automatically dismissed. Typically, distilled water, from which all salts and minerals have been removed, is used in research. With the polywater case, it was generally concluded that the unusual structure and properties of water were due to impurities in the water.

~

However, the greatest damage to water research was caused by the media reporting on the case. It was suggested, among other things, that a drop of such polywater thrown into the ocean could act in the same way as other polymer catalysts – it could initiate a polymerization process, and all of the Earth's water resources could suddenly turn into one giant polywater ball, eventually leading to the cessation of all life on Earth. Of course, this was understood as a joke, but in any case, water researchers were portrayed as incompetent and were belittled. Nowadays, such behavior on social media seems to be all too common, and the event might not receive publicity in the same way. At that time though, the impact of negative publicity was really detrimental to water research — it went through several quieter years. Many esteemed researchers who could have conducted water research chose safer topics instead, probably out of fear of being discredited. A couple of decades later, just as the polywater case had been forgotten, water research experienced another shocking case that many remember even better.

~

Benveniste and the memory of water

In 1988, French researcher and esteemed immunologist Jacques Benveniste, along with his colleagues, obtained evidence that water could retain information from molecules with which it had interacted. Water seemed to have the ability to "remember."

Benveniste and his team studied solutions of biologically active substances. They repeatedly diluted these solutions. Each time, they took the diluted solution and diluted it further, until nothing was left but pure water. Statistically, none of the original substance should have remained. They diluted the solutions even beyond this point, where really nothing should have been left. However, the solution seemed to have as much effect as the original undiluted solution. The same effect was achieved by pouring onto cells both the original and the several times diluted solution. The diluted solution, or rather just plain water, seemed to retain a "memory" of the molecules with which it had interacted. Only those biologically active molecules originally in the water could have an effect in the way that plain water now did.

Benveniste submitted a publication of his studies to the prestigious journal *Nature*, whose editor-in-chief Sir John Maddox dismissed Benveniste's research. He considered it absurd to think that water could retain information. After receiving a rejection letter for his findings from *Nature*, Benveniste asked scientists from three other laboratories to repeat what he had done in his experiments, to see if they could achieve the same results. And indeed, they did. Benveniste sent a report of the new experiments to *Nature*, and eventually, the journal agreed to publish the results, but with one peculiar condition: The editor wanted to assemble a committee that would oversee Benveniste's team of researchers as they conducted their experiments.

The committee would then report its findings to the readers of *Nature*. Benveniste's group ultimately accepted this unusual requirement. An article on Benveniste's research appeared in *Nature*, accompanied by a disclaimer from the editor: the editor mentioned wanting an investigation where a group of peer reviewers would observe the conducting of the experiment.

However, it seemed that the researchers were suspected of more than just a minor insignificant mistake, as the composition of the committee was interesting. In addition to editor-in-chief Maddox, the committee included Walter Stewart (US National Institutes of Health), who specialized in uncovering scientific fraud, and magician James Randi, also known as "The Amazing Randi." World-class stage magician Randi was known for revealing the secrets behind other magicians' tricks. The committee's purpose appeared to be solely to discredit Benveniste's team's work and to demonstrate that they were wrong.

The committee arrived in Paris to observe the experiments. The first series of tests went as reported. The situation looked good for Benveniste's team. However, when one of the committee members wanted to make the dilutions independently, the results were no longer as good. The committee concluded that since the researchers themselves were able to produce the results, but the committee members were not, some sort of trick must inevitably be involved. However, the nature of the supposed trick remained unexplained. Despite everything, they reported to the scientific world that the memory of water was an illusion.

Benveniste later responded to the accusations in *Nature* in the fall of 1988, stating that in their publication, they presented things as they had documented them. Their publication was a call for help to others to explain these peculiar results. Instead, they received accusations of fraud. Benveniste wondered in his response how all – by then, five –laboratories that had replicated Benveniste's results could also have been involved in the fraud? It is unnecessary to go into the details of this controversy, as you have likely already grasped the situation.

In any case, as a result of the fiasco, Benveniste experienced widespread humiliation. He lost his grants, his laboratory, and faced difficulties in publishing any scientific work. Moreover, he was awarded the satirical Ig Nobel Prize twice, which Harvard students give annually for questionable research. The name of the award comes from the word *ignoble*. However, Benveniste did not give up. He continued his research until his death, funding it himself.

Something that indicates the attitude of other researchers is that upon Benveniste's death in 2004, *Nature* published an obituary in which the author stated that Benveniste's work cannot be judged solely on the basis of the second half of his life. In addition to the questionable research, one should also remember his "significant" research from the early part of his career.

~

With these cases, it's easy to understand the divided state of water research. Decades have now passed since Benveniste's case. The situation has gradually improved over the years. Yet, water research still carries the stigma of pseudoscience to some extent, and few researchers are enticed by a research topic for which it is difficult to find funding. Many researchers begin to pursue more open-minded research as a sort of a hobby after their formal research careers have ended, when their own careers are no longer at stake as much. There are, of course, pioneers in water research, and I discuss many of them in this book.

Water myths – stories and beliefs – speak of water's central role throughout the ages. Water has been perceived as a foundation that existed before everything else. Myths were born from the need to understand and explain the world. This need still exists, and understanding is also sought through science and research. So, what is currently known about water?

CHAPTER 3. WHAT IS WATER?

"What we know is a drop,
what we don't know is an ocean."

– Isaac Newton

Water is so commonplace that many of its properties would seem self-evident. However, I will start with the basics, as we might overlook something essential, thinking we already know water well enough. The researcher's duty is to approach any subject or mystery with an open mind, as if encountering the topic for the very first time.

When examined through the senses, water is transparent, almost colorless, and usually odorless and tasteless. It has no shape; it takes the shape of its container. It has no sound nor its own movement, but it can be moved by the wind, as it puts up little resistance. Water's movement can be minimal, in ripples, or it can form gigantic waves or tsunamis that can destroy shores and roll far inland.

Water appears in solid form as ice, in liquid form as water, and water vapor or steam. The gaseous form of water is invisible and is practically everywhere.

WATER = H_2O

The water molecule itself is quite well known. The structure of water was apparently first determined by the London researcher Henry Cavendish around the year 1781. Sometimes, the Scottish inventor James Watt, who published his own research in 1783, or the French physicist and chemist Joseph Louis Gay-Lussac in collaboration with the German researcher Alexander von Humboldt (1805), are also presented as the "discoverer" of water.

Cavendish was an introverted researcher who avoided social contact and feared people's opinions. Many of his studies were published only after his death. Cavendish had conducted many studies and discovered phenomena that were credited to other researchers, such as Ohm's law, Dalton's law of partial pressures, or Charles's law of volumes. Biologist Rupert Sheldrake explains the phenomenon of concurrent discoveries through the concept of morphic fields: we are in the same information field, so it is not uncommon for almost the same discovery to be made at the same time in different places.

In any case, Cavendish correctly predicted the composition of water in his research: two parts hydrogen combined with one part oxygen to form a water molecule.

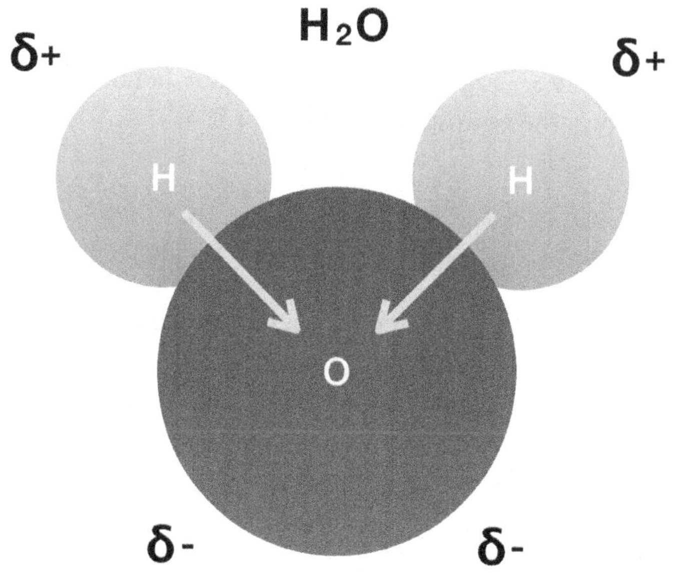

Figure 2. H_2O molecule where two hydrogens combine with one oxygen atom O. The δ+ and δ- represent the partial negative or positive charges in the molecule.

This image (Figure 2.) is, of course, just a model of water. In reality, water is not formed from such "Mickey Mouse" figures, of which the partial charges within the molecule (δ+ and δ-) remind us. Although water is electrically neutral on the outside, it behaves like a small magnet (an electric dipole): the oxygen atom attracts the shared electrons of the oxygen-hydrogen bond more strongly, which is why the oxygen atom acquires a partial negative charge (δ-), and hydrogen, in turn, a positive partial charge (δ+). Due to the two electrons missing from the outermost electron shell of the oxygen molecule, the hydrogen atoms are not symmetrically arranged around the oxygen, but the molecule bends so that the angle between the H-O-H atoms is about 104.5–106°. The bond between oxygen and hydrogen is about 0.0965 nm in length.

Many of water's unique properties are explained precisely by the polarity existing between the water molecules that form it – this is referred to as the dipolar nature of water. Water molecules form hydrogen bonds with other water molecules, but also with other polar molecules. In pure water, water molecules are on average bonded by 3.4 hydrogen bonds to other water molecules, while in ice, the corresponding number is 4.

WATER = A NETWORK OF MOLECULES LINKED TOGETHER BY HYDROGEN BONDS

Water molecules are interconnected through hydrogen bonds. A hydrogen bond forms between a hydrogen atom (H) and the free electron pair of an oxygen atom (O) from another water molecule. Hydrogen bonds can only form between hydrogen and another electronegative atom, which means an atom with a negative charge. Water molecules arrange themselves relative to each other in a way that allows hydrogen bonds to form optimally, with the least amount of energy. The hydrogen bond is, in fact, an electrostatic attraction between atoms, about ten times weaker than covalent bonds but stronger than the interaction between two dipoles, for example. Since hydrogen bonds are not very strong, especially in liquid water, they constantly form and break between water molecules. Water also acts as an excellent solvent; it dissolves almost everything. One reason for this is water's ability to form electrostatic interactions with other polar molecules and ions, due to its own polarity. Water dissolves ions that carry an electrical charge, such as salts, acids, and bases, particularly well. When heated, water's solubility often increases even more. You have probably noticed sugar dissolves better in hot coffee than in cold.

Water molecules form a network interconnected by hydrogen bonds. Hydrogen bonds can explain some of the properties of the structure of water, but there is still much to be studied. For example, one consequence of hydrogen bonds is that water in its liquid state is denser than in its solid state.

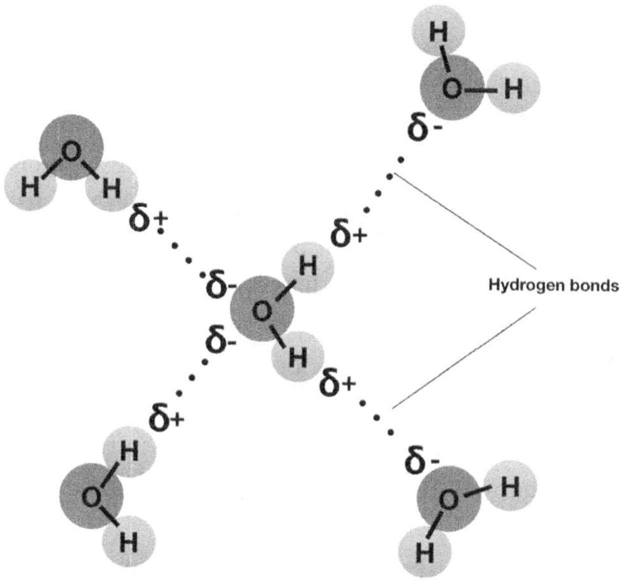

Figure 3. Water meaning hydrogen bonds between water molecules.

~

Ice floats on the surface of water and not vice versa

As I write this, there is a slight frost outside and half a meter of snow. It's December, and the sea is not yet frozen here in Helsinki, Finland. Have you ever wondered how important it is that ice floats on the surface of water, rather than sinking to the bottom?

If ice were heavier than liquid water, it would sink to the bottom of the sea or lake, and freezing would start from the bottom. As a result, lakes would eventually freeze throughout their entire depth. The bottom layer of ice might not even melt completely during summer, as the sun-warmed water would stay on the surface. Therefore, there would be no life in lakes, and it is very likely that there would be no life at all in the northern latitudes.

~

Why then does ice float? Water is densest at 39 °F (+4 °C). Usually, the density of liquids only increases as the temperature decreases, all the way down to the freezing point. The reason water is densest precisely at 39 °F is due to the packing of water molecules caused by the hydrogen bonds between them. The density of water at 39 °F is 8.345 pounds per gallon (1 kg per liter). Water's density is greater than that of ice, which is 0.917 kg per liter. The structure of ice has more empty space due to the crystal structure formed by hydrogen bonds (Figure 4). Because of its lower density, ice floats on the surface of water. Water sinks to the bottom at 39 °F, while surface water tends to follow the air temperatures. In winter, the warmer water is at the bottom, and in summer, the opposite is true. As it melts, the density of ice

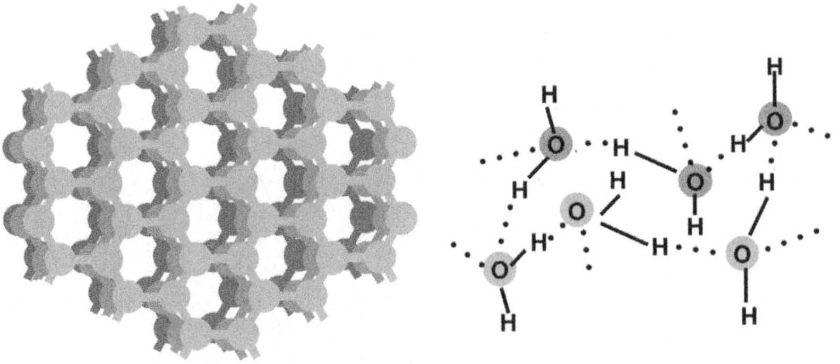

Figure 4. In ice, water molecules are organized into a crystalline structure. On the left, only the network of oxygen atoms is depicted, while on the right, the placement of hydrogens is shown. The molecules are interconnected by hydrogen bonds.

increases when the crystal system loosens and breaks down, collapsing. At temperatures above 39 °F, water behaves like other liquids, that is, it expands when heated (density decreases). In spring and autumn, the waters undergo a vital vertical circulation between the bottom and surface layers as the temperature either warms up or cools down.

Although the hydrogen bond is cited as the reason for water's anomalous properties, its role is not yet fully understood. Linus Pauling, who won the Nobel Prize in Chemistry in 1954 for his research on chemical bonds and molecular structures, proposed in the 1930s that the weak hydrogen bonds in water mimic their behavior from the stronger covalent bonds of water, thereby reinforcing them.

Pauling assumed that this must occur because, in water, electrons and bonds also follow the strange laws of quantum mechanics. All objects – even solids – can behave like waves under certain conditions. In water, this means that the wave functions describing the behavior of electrons in covalent and hydrogen bonds can overlap, thereby transmitting information. This phenomenon is called delocalization, where an electron does not belong to any specific bond or atom but collectively to the entire molecule. Delocalization usually means that a molecule is more stable than without delocalization. Electrons cannot be seen as separate entities, but they share information among themselves. Pauling's proposal was controversial at the time, but at the turn of the 21st century, a US-French-Canadian research team (E.D. Isaacs, et al.) confirmed the hypothesis in their X-ray diffraction experiments: this indeed happens. The behavior of hydrogen bonds mimics that of covalent bonds. The research team estimated that about 10% of the behavior of the hydrogen bond is due to the effect of covalent bonds.

It appears that water molecules reinforce each other. Indeed, in his hypothesis of morphic fields, Rupert Sheldrake proposes that nature consists of such wholes, which are more than just the sum of their parts. Every self-organizing system is a whole formed from its parts, whose properties are determined at each level by the morphic field. Also crystals and molecules have their characteristic crystal and molecular fields at these levels.

WEAK BUT MEANINGFUL CONNECTIONS OF WATER

We have only recently begun to understand how significant a role the unique molecular connections play in creating the properties of water.

~

The melting and boiling points of water are exceptionally high

A hydrogen bond can form with oxygen, nitrogen (N), or fluorine (F) atoms, so hydrogen bonds are also present not only in water, but also in substances like ammonia (NH_3) and hydrogen fluoride (HF). When in solid or liquid form, water and these substances form clusters consisting of multiple molecules due to hydrogen bonding. This makes the melting and boiling points of these compounds much higher than those of other hydrogen compounds that do not have hydrogen bonds.

The effect of the hydrogen bond, especially in water, is surprisingly large if, for example, it is compared to the melting and boiling points of other dihydrides (H_2Te, H_2Se, H_2S, and H_2O, all of which have two hydrogen atoms) that belong to same group 6A of the periodic table as oxygen (*Table 1*).

GROUP	3A	4A	5A	6A	7A
	B_2H_6: -92,5	CH_4: -164	NH_3: -32,4	**H_2O: +100**	HF: –87,7
				H_2S: -60,7	
				H_2Se: -41,5	
				H_2Te: -2	

Table 1. The boiling points (°C) of hydrides of elements close to oxygen in the periodic table. [P. Wiggins]

When compared to other dihydrides, the boiling point of water should be around -65 °C. The actual boiling point of water is therefore about 165 °C higher than one might expect! Similarly, the melting point of water is about 90 °C higher than one might anticipate. And fortunately this is the case, for otherwise water would have vaporized away from the Earth long ago, and we would not exist!

The anomalous boiling and melting points of water cannot be explained by hydrogen bonds alone. It seems that sometimes the effect of hydrogen bonds is even further intensified. In addition to copying the properties of covalent bonds (when an electron is shared), the effect of the hydrogen bond can become even stronger. Water molecules are actually capable of forming weak bonds over a very long distance. This is due to the dipole nature of water: water molecules attach to each other not only through hydrogen bonds but also through several weaker bonds. These weaker bonds are called long-range interactions, non-bonded interactions, or also van der Waals forces, named after the Dutch physicist Johannes Diderik van der Waals. He received the Nobel Prize in Physics in 1910 for his studies on gases and water. Another physicist, James Clerk Maxwell, noted upon the publication of van der Waals' first scientific paper: *"There is no doubt that 'van der Waals' will soon be among the leading names in molecular physics."*

Weak Connections Strengthen Water's Properties

- Coulomb interactions, which are due to electrical charges, are often also considered part of van der Waals (vdW) interactions. Because of charged ions – there are always small amounts of them in water – water conducts electricity.

• Water also forms ions itself, protons H^+ and hydroxide ions OH^-. The concentration of H^+ ions (or actually H_3O^+) is considered a measure of acidity.

• vdW interactions include a long-range attractive effect, pulling particles together, as well as a short-range repulsive effect, pushing them apart. Thus, atoms repel each other when they are very close and attract each other when they are further apart.

T. Morawietz and others describe in their publication "How van der Waals interactions determine the unique properties of water" (2016) how vdW bonds are particularly significant in forming the properties of water. They use a model based on neural networks in their calculations, which takes into account long-range interactions. In this way, the dynamic properties of water can be described more realistically. According to simulation calculations, vdW bonds seem to further strengthen and in some cases weaken the effect of hydrogen bonds.

As a result of the combined effects of hydrogen and vdW bonds, water is able to change its structure and thereby its properties in an exceptional manner. Various changes, such as those in the environment, cause water to react sensitively. Weak bonds are prone to change, allowing water to alter its structure. The structure also determines the properties, and it is thus easier to change that than, for example, its composition. A good example of this is carbon, in its forms diamond and graphite. Both consist of carbon, but diamond is one of the hardest substances on Earth, while graphite is one of the softest.

TWO KINDS OF ICE, TWO KINDS OF WATER

Ice melts at 32 °F (0 °C), and water vaporizes at 212 °F (100 °C). The Celsius scale is actually based on the phase transition points of water. The Swedish scholar Anders Celsius initially named the unit in 1742 as the centigrade (from Latin centum, "hundred," and gradus, "steps") and the scale was the reverse with the boiling point at 0 and the melting point at 100, to avoid negative numbers. However, the scale was later inverted. In some countries, including the USA, the Fahrenheit scale is used. In the scientific community, Kelvin degrees are used, where 0 K = -459.67 °F = -273.15 °C, which is also the absolute zero point and the point at which atomic thermal motion ceases.

The structure of a molecule affects its properties. Water, even in different states of matter, has several different forms.

Ice has several stable structures

Unlike other materials, ice – solid water – has several different forms. In physics classes, it is taught that water can exist in three different states of matter: solid (ice), liquid (water), and gas (water vapor). The H_2O molecule can also exist in more than one solid form. In other words, there is more than one type of "ice."

Water turns solid at 32 °F. The solid ice familiar to us is a crystalline material where molecules are organized into a lattice structure. In addition to this, there is also amorphous ice, whose structure is not as regular. Amorphous ice lacks long-range order. This form of ice exists in two types, high-density and low-density forms.

Amorphous ice is not snow. Snowflakes are actually ice crystals that form when water vapor condenses at temperatures below the freezing point. The lower the temperature, the smaller the snowflakes that form. Thus, ice exists in two external forms: solid ice and snow formed from small ice flakes. Solid ice exists both in an organized form (regular ice) and an unorganized (amorphous) form.

Amorphous ice forms when water is rapidly cooled to very low temperatures. When this ice is then heated at 130 K (in Kelvin) or about -220 °F (-140 °C), the amorphous ice turns into crystalline form. In a joint study by researchers at Princeton University and the City University of New York (F. Martelli et al.), computational simulations were used to show that water molecules in amorphous ice organize into unpredictable forms that were not present in the original liquid form. Previous understanding suggested that such rapidly frozen "glassy" structures "inherit" the structure of the former liquid form. In the case of amorphous ice, it now appears that entirely new forms of organized molecules are formed. The structure is not the same as in the case of ordinary, well-organized crystalline ice. Amorphous ice is not common on Earth, but it is the form of water that, according to studies, is most abundant in the universe (C. Zandonella).

~

The existence of two different solid forms of ice could imply that there are also two different types of water. Experimentally, it has been observed that at room temperature, water appears either in a low-density form with a tetrahedral structure or in a high-density form. Indeed, over the years, many researchers have discussed the two forms of water.

The discoverer of X-rays, William Röntgen, in his 1892 publication "Concerning the structure of liquid water," suggested that water has two structural types: ice-like organized and liquid-like not-so-organized water. In one state, organization is greater, and hydrogen bonds form maximally, which limits the number of neighboring molecules. In the other state, disorder is more characteristic, allowing molecules to pack more tightly and allowing fewer hydrogen bonds.

The model of two types of water and water's "crystalline, gel-like, or organized" state has been discussed in many studies since then. Sir William Hardy proposed as early as 1912 that we should talk about a fourth state of water. Dr. Gerald Pollack was the first to propose the molecular structure H_3O_2 for this organized form and hypothesized about the role of this fourth state in water's behavior, emphasizing its impact on human health as well.

Altering the structure is particularly significant. In her article "Life Depends Upon Two Kinds of Water" (2008), Professor Philippa Wiggins mentions that if water, even locally, can change its density, this implies changes in hydrogen bonding and, consequently, in water's physical and chemical properties. Naturally, this also means changes in vdW bonds, which are more sensitive than hydrogen bonds. Wiggins spent the majority of her life, over five decades, studying the properties of water and published her findings on the two different forms of water as early as 1986. Her research has had a significant impact on the scientific understanding of water, although unfortunately, it has not received as much publicity as it deserves.

It appears that the anomalous properties of water are the result of the interactions between water molecules and how molecules organize under different conditions. Is water then truly anomalous? It is presented as such, but in reality, our current models are simply incapable of accurately describing water.

ABOUT THE QUANTUM EFFECTS OF WATER

Quantum physics describes the behavior of atoms and particles, and when examining water, one should indeed take into account quantum phenomena, as Linus Pauling noted in the 1930s. Water should be considered a quantum mechanical molecule, especially because it contains two hydrogen atoms, which do not have a large mass. The smaller the masses, the more quantum phenomena need to be considered. In quantum physics, phenomena are described using waves. For example, the state of a particle is described using probability waves: it is possible only to present probabilities, for instance, of how likely a particle is to be found in a specific place. In larger molecules, the significance of individual hydrogen atoms is not as crucial for the whole, and they can be described with sufficient accuracy using classical theory and classical (Newtonian) physics without needing to specifically account for quantum phenomena.

Increasingly, researchers believe that quantum phenomena should be taken into account to both understand water and interpret experimental results. Researchers Kim Hyeon-Deuk and Koji Ando from Kyoto University are of the opinion that quantum phenomena should be considered when computationally modeling the hydrogen-bonded network, in other words, water and its anomalous properties.

In recent years, an increasing number of studies of water have been done using computational methods, as computers and methodologies have significantly advanced. Water has been particularly challenging to model because of the necessity of accurately depicting long-range interactions to replicate its true behavior. Typically, these interactions have been omitted from calculations because it was thought they were not significant. However, if the model created does not reflect reality, then it is pointless to use it. Often, simplifications must be made in molecular modeling because, even though the modeling is done on supercomputers, the computation time increases with the size of the system being studied and the number of interactions that need to be considered. However, the situation is quite different from when I was conducting my doctoral research over 20 years ago. Computing power has greatly increased. At that time, the world's most powerful computer, IBM's ASCI White, had a peak computing speed of 7.226 Tflops (10^{12} floating-point operations per second), while currently (July 2024), ordinary laptops may reach the same class, and the fastest supercomputer in the world, Frontier (HPE Cray EX system) has achieved a performance of 1.206 Exaflop/s (10^{18}), which means it can perform incredible over a quintillion (1 followed by 18 zeros) calculations per second.

Obtaining results takes longer the more accurately one wants to depict all the properties of water. The better a model accounts for the various interactions between water molecules the more precise it is. Vega (2011) describes the relationship according to the following list. The number following each model represents the relative computational time and is only indicative due to the development of computational powers:

1. Rigid ball-and-stick model = 1 (as shown in Figure 2, atoms are represented by balls, and the bonds between them are depicted as sticks)
2. Flexible ball-and-stick model = 5
3. Rigid ball-and-stick + model taking into account water polarization = 6
4. Flexible ball-and-stick + model taking into account water polarization = 30
5. Flexible ball-and-stick + model taking into account nuclear quantum effects = 175
6. Rigid classical nucleus + electrons' quantum behavior through computational density functional theory (DFT) = 1000
7. Flexible classical nucleus + DFT = 5000
8. DFT + flexible nucleus + nuclear quantum effects = 175000

Therefore, if a simulation with a rigid ball-and-stick model takes 1 hour, it would take 175,000 hours or about 20 years with a model that takes into account quantum phenomena! Models 1–4 are based on classical physics, whereas models 5–8 take into account quantum phenomena. The more precise the model, the more challenging it is to handle computationally.

WHAT KIND OF A MODEL WOULD DESCRIBE WATER BEST?

How then can water, its interactions, and properties be accurately depicted in different cases? It must be noted that more anomalous properties of water are continually being discovered as we learn more about it. A model that perfectly describes water does not exist yet.

In any case, the simple "Mickey Mouse" model often does not suffice, and there are many alternative, more complex models. Even experts in the field often have difficulty explaining differences between them, and I will not delve into them here in detail. Generally, it can be said that current models of water aim to depict its behavior and, among other features, the following aspects of water:

- **Clusters.** The tendency of water molecules to bind together into clusters through hydrogen bonds, among other means. Frank and Wen's (1957) "Flickering clusters" model describes water in this way. In their model, clusters have a critical size after which they break apart. The formation and breaking of clusters occur very rapidly – hence the term "flickering."
- **Multiple states of liquid water.** Many models suggest that liquid water could have multiple states. For example, Martin Chaplin (London South Bank University, 1990) has proposed that liquid water consists of two types of intermixing nanoclusters, in which water molecules rapidly change states. Chaplin has focused particularly on intracellular water in his research.
- Likewise, Lars Pettersson (Stockholm University) and Anders Nilsson (Stanford University) have proposed a model with two distinct types of water: an ice-like organized part forming chains or clusters and a disorganized part surrounding the clusters. *"New results strongly support the view that at room temperature, water cannot decide which form it should be, high or low density, leading to local fluctuations between the two,"* Pettersson states. Depending on the environment, these different forms manifest in water in varying proportions. They similarly form two types of amorphous ice at very low temperatures upon freezing. Observations suggest that water is in fact in a constant state of change between its denser and less dense forms. Depending on conditions, such as temperature and pressure, water

favors one form over the other. According to Pettersson, water is not a complex liquid but two simple liquids with a complex relationship.

- Sergey Pershin (Russia) and Yosi Scolnik's (Israel) model posits that water has two separate forms that are mirror images: left-handed and right-handed forms. According to the model's developers, these different forms and their relationship could explain water's exceptional properties. In Professor Rustum Roy's (Pennsylvania State University) model, water is described with a heterogeneous structure, taking into account the interactions within water and between water and dissolved molecules.
- **Quantum Phenomena in Water.** Del Giudice (University of Milan, 2013) describes water using a cluster model based on quantum field theory. The smaller the mass of a molecule or atom, the more quantum theory should be taken into account.
- **Anomalous Properties of Water.** Among others, John Russo (University of Bristol, 2018) aimed to replicate the behavior of water and thus its unique properties using computational methods. At around 80 °F (27 °C), approximately 90% of water molecules are bound to each other by hydrogen bonds. This result has been reached through simulations done on supercomputers as well as experimental measurements. In the network of molecules formed by hydrogen bonds, the hydrogen bonds continuously break and reform, with their duration being about 10 fs (femtoseconds, 10^{-15} s) to approximately 10 ps (picoseconds, 10^{-12} s).

WATER = LIFE

It's astonishing how exceptional water is. If water did not behave as anomalously as it does, we would not exist. Water is everywhere on Earth and is vital for us: the majority of the human body is composed of it, it is in our food, and in the air we breathe.

It was long thought that water formed well after the universe was created. Astronomers believed that all molecules heavier than hydrogen were formed later in the cores of stars and that it took time before hydrogen and oxygen met. Hydrogen is the most common element in the universe, and oxygen is the third most common. From this, therefore, one can infer that water is fairly common in the universe. A study published in 2015 by Tel Aviv University and Harvard University suggested that water was present in the universe much earlier than previously thought, when the universe was about 5% of its current age. The timing of water's appearance is crucial for life, as current knowledge suggests that water is necessary for the emergence of life, and the first planets to form could already have had conditions suitable for life.

Our understanding is that water is necessary for the origin of life. When searching for life elsewhere in space, water is the first thing we look for. In fact, wherever there is water on Earth, there is also life. Overall, about 70% of Earth's surface is covered with water – the same percentage as the amount of water in our human bodies. Arthur C. Clarke, the screenwriter of "2001: A Space Odyssey" and an oceanographer, aptly remarked that a much more fitting name for planet Earth would be Ocean. The majority of Earth's water is salty seawater, while on the continents, water is fresh, so-called freshwater. In fact, freshwater is quite scarce, only about 2.5% of Earth's water, and of this, approximately 70% is in glaciers. Therefore, up to 96.5% of Earth's water is in the seas.

When I started primary school in the 1980s, water had not yet been discovered on Mars. That planet's conditions were considered impossible for water, and it was thought to be arid. However, it was speculated that since there were features on the surface resembling riverbeds, water had once been present there. These surface formations on Mars have inspired many, including H.G. Wells's fantasy book *The War of the Worlds*. The book was very popular when it was published in 1897. In the book, Martians attempt to conquer Earth after Mars became uninhabitable. The book also seemed to anticipate the upcoming First World War.

Through infrared imaging, scientists have discovered that there is still water on Mars, even in liquid form and flowing. Mars appears to have much more water than previously thought. Radar measurements have found water beneath Mars's southern polar ice cap and even entire subglacial lakes. The water on Mars is thought to be highly saline, as scientists have found streaks of water that are dark when the temperature is above -17 °F (-27 °C); below this temperature, they disappear, presumably freezing. The presence of salts lowers the freezing point of water. This same phenomenon is why roads are salted to keep them drivable as temperatures approach 32 °F.

It is now also known that there is water on the Moon. Towards the end of 2020, *Nature Astronomy* published several studies on lunar water. For the first time, infrared imaging indisputably showed the presence of water in molecular form on the Moon's surface. Water has been found both at the lunar poles as ice and elsewhere bound within the Moon's mineral soil. However, utilizing this water is not easy in its current form. Moreover, the water is extremely scarce, even more so than in the sands of the Sahara.

We now know that in our solar system, water exists on many planets and in their atmospheres and compositions. The four largest planets – Jupiter, Saturn, Uranus, and Neptune – appear to contain vast amounts of water, as do their moons and the rings that surround these planets. Five of the moons of Jupiter and Saturn seem to be made up entirely of oceans: Jupiter's Ganymede, Europa, and Callisto, and Saturn's Enceladus and Titan. According to images taken by NASA's Hubble Telescope, the water on Ganymede appears to be saline, with liquid water existing beneath a frozen surface. Europa and Enceladus also have subterranean liquid seawater. Enceladus appears to have hydrothermal activity at the bottom of its oceans, meaning warm water. This environment could be suitable for living organisms.

Astronomers have also observed signs of water in gigantic molecular clouds surrounding stars, in protoplanetary disks where new stars are formed, and in the atmospheres of giant planets orbiting stars. Water seems to be everywhere in the universe, providing the conditions necessary for life as we know it. Of course, life could theoretically be defined chemically in other ways, and water could potentially be replaced by something else. Ammonia is one such alternative, as it is liquid from about -130 °F (-90 °C) to 140 °F (+60 °C). However, being in a liquid state alone does not make conditions Earth-like.

Water is essential for living organisms precisely because of its unique properties – how it acts as a solvent, its ability to bind heat, and the fact that it is liquid at human body temperature. Water plays a significant role in the processes of living beings. It interacts with various biological molecules, including proteins, DNA, and also acids, salts, and sugars. However, too little is still known about these interactions and the interactions between water molecules themselves. Albert Szent-Györgyi, often described as the father of modern biochemistry, characterizes the indispensability of water as follows: *"Life is water dancing to the tune of solids. Without that dance, there could be no life."*

~

Water and life take my thoughts to the birth of my first child. There were still two weeks left until the due date when water, meaning the amniotic fluid, suddenly broke. While we were test-driving a car that day, I felt a very firm kick to my insides in one bend of the road. I thought already then that perhaps it was time for the birth. Later that evening, while lying on the sofa, my waters broke. It felt frightening; my baby no longer had the surrounding protective layer of water. I was also prepared to wait another month because often, especially with one's first child, one has to wait well beyond the due date. We called the hospital. They believed that I would give birth within a day. I was instructed to wait a little while longer at home and rest, as contractions had not yet started. Soon they began, and by four in the morning, we checked into the Maternity Hospital. I had hoped for a water birthing room, and to my delight, I got one. I had imagined it to be something of a spa experience: I would soak in a warm pool with bubbles, and my husband would massage me while my child gently emerged from the womb and swam up into my arms.

For a while, it truly was like being in a spa, but unfortunately, the contractions disrupted the tranquility. I don't remember if it was because the waters had broken or because it was my first child, but at some point, the midwife wanted me out of the water for an examination. After the examination, the contractions became so frequent that I no longer wanted to return to the pool.

I played a recording that I had listened to throughout the pregnancy: a relaxation exercise with the sound of ocean waves. The contractions came in like waves too, leading up to the birth of my child. The midwife described the pain as the crest of a wave: it's best not to resist but to let the contraction do its work and ride over the pain on the crest of the wave. In the morning, I held my little one in my arms, who is, fittingly, an Aquarius.

From water we are born, and it is an essential part of life. According to water researcher Dr. Mae-Wan Ho, it is specifically water's quantum coherence that enables life. In the case of water, quantum coherence refers to how information travels within it – how close two states are to each other. For example, two waves are in perfect coherence with each other when their wavelength and frequency are exactly the same (Figure 5.). Coherent waves also amplify or diminish each other, that is, they interfere with each other.

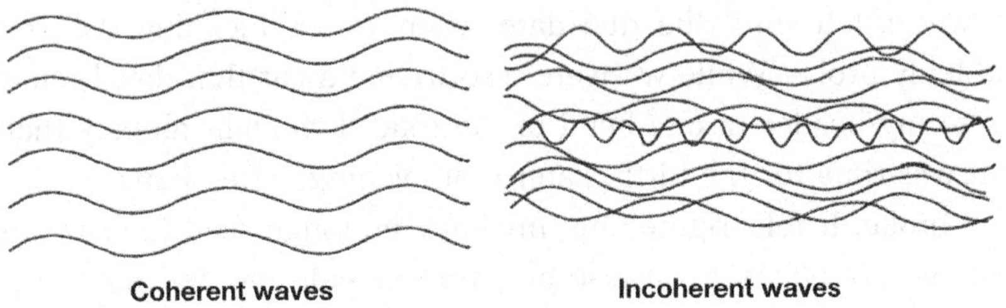

Figure 5. Coherent waves on the left, incoherent waves on the right

~

According to the term's originator, John Stewart Bell, quantum coherence allows for faster information transmission even across distances. Quantum coherence is known to occur for electrons, photons, atoms, molecules, and even macroscopic objects, and this is also exploited in the development of quantum computers.

Water is a sensitive liquid, and it must be examined at all possible levels. The essence of water cannot be captured by studying it solely in a laboratory. We must not forget that we are water. Water must be given the opportunity to reveal its character in its natural environment and in ways that are inherent to it. Therefore, water must also be approached experientially, by seeking a connection with it.

CHAPTER 4. COLD EXPOSURE

"In front of the mystery of water and the mountain summit's thrust,
male supremacy is abolished;
walking through the heather, dipping her hand in the river,
she lives not for others but for herself.
The woman who maintained her independence through all her
servitudes will ardently love her own freedom in Nature."

– Simone de Beauvoir, The Second Sex, part II, p. 744

I received this quote in a message on my phone from a friend just a few days after I had decided that my new book would be tentatively titled *The Mystery of Water.* My friend did not know about this. I have experienced similar synchronicities while writing previous books as well. At this time, I had just started cold water swimming. I used to swim in a group that met regularly for sauna and winter swimming sessions. However, the group stopped meeting in the fall of 2020 due to meeting restrictions, as communal saunas were not allowed (COVID-19). One phrase from the quote, "male supremacy," symbolizes societal authorities and sometimes quite absurd rules for me. The flow and freedom reminded me of water, which also symbolically purifies, resets, grounds, and brings energy.

In this chapter, I discuss my own experiences with cold water swimming. Please note, I am not a medical professional. Before starting cold water swimming, one should determine if it is suitable for oneself.

WATER = FREEDOM

It's the first of January. For Christmas, I received Marjut Hjelt's book *Maaginen vuosi* which in English means *The Magical Year*. It described the old belief that the first day of the new year would mirror the whole year, based on one of the most important rules of superstition that *like causes like*.

Determined not to take any risks with how the year would progress, I decided to run an 8-kilometer loop and take a dip in an ice hole.

It was just a few degrees below freezing, and running felt comfortable. As the first day of January was turning into evening, I saw a reddish-orange twilight on the horizon. I had just joined a local winter swimming club that maintains an ice hole at the nearby beach and a heated container for changing clothes. Going swimming from the container felt luxurious, as previously, I would head straight to the sea from the beach after my run. I wore my swimsuit under my running clothes to make going for a swim easier.

Swimming in cold water feels like a sacred event. It closely resembles the initiation and activation ceremonies found in spiritual traditions. I've studied Japanese Reiki, a method for internal growth and natural well-being. In this tradition, following a certain formula or path leads to hidden, higher knowledge, and mastery of this knowledge requires activation through initiation. A seasoned master or teacher can assist in this. The teacher takes the shape that is needed for the student. I feel that water is now my teacher, and my task is to discover who I really am and what my deepest essence is. Looking deeply is also a form of surrender. When one dares to surrender and let go, a path unfolds, becoming increasingly clear.

When I go into the container to leave my running clothes there, it's like entering some kind of portal, a sacristy, before stepping into the holiest of holies. I leave the clothes hanging on a rack, silence prevailing. There are no others in the container. I write my name in the container users' guestbook. The time is 3:15pm. I step out into the frost in a bikini, barefoot. The cold wind blows on my skin, still sweaty from the run. The corner of the container is slippery with ice; I must be careful to stay upright. I walk to the pier. A couple of passersby walk past the shore in their warm clothes, walking a dog. It feels odd that I am toddling around in a bikini. However, I feel empowered. I am proud to be here and doing this thing that seems odd to some right now. I remember the sentence: *"Only dead fish go with the flow."* One must do exactly as feels right. After the run, adrenaline surges, and nothing can stop me.

A heating mat has just been installed on the pier. It feels like walking down a red carpet, although the mat is not red but blue. I walk to the end of the pier to the stairs, take one deep breath in and then, as I breathe out, I

Figure 6. There is a dot in each intersection but of which color?

immediately descend the cold, steel stairs into the sea. I count down in my mind 10 - 9 - 8 - 7, and so on, all the way to 1 as I immerse myself in the water. My hands almost freeze to the railing of the stairs. I never stay to test how cold the water is. The water does not turn to warm; it is supposed to be cold. I submerge myself in the water and do a short swim, about fifty strokes, before I stop to look at the scenery from the water. I feel one with the water and the landscape, the sense of separateness disappears. I do not feel cold because I do not focus on it. I feel how warm blood flows inside me. The best advice for going into cold water I heard from my daughter. She told me to focus on where it is warm. When you bring your attention inward, the warmth of the heart helps. At the same time, gratitude spreads everywhere.

I feel how the water "speaks" to me. It grounds, resets. The water refreshes, and I feel how it activates, increases life energy. I stay in the water for about two to five minutes in the winter, depending on whether I am alone or not. The experience is sacred, and I prefer to be there alone. I believe many others do as well. I often notice how someone stays stretching on the pier for a while, probably just waiting for the other to come out of the water. The activation by cold water must be experienced alone.

The most challenging phase is returning from the ice hole to the container. Today, it's so windy that my thighs and buttocks freeze immediately upon getting out of the water, almost completely losing sensation. But as I walk towards the container I feel alive! Inside, I try to rub warmth into my thighs and buttocks, even though I know they will warm up as I run the remaining mile or two home. This all may sound crazy, but the feeling afterwards is so amazing!

There is something so special about swimming in cold water. I can decide to do differently from what my senses initially tell me. The senses do not always tell the truth, and in fact, they often lie. Or actually, the brain interprets the messages from the senses, fills in missing information, and sometimes even provides incorrect information. For example, look at Figure 6: What color does the dot at the intersection appear to be? When you do not focus your gaze on it, it no longer seems white.

However, I have the freedom to choose. We may have learned that our senses tell us the water is too cold, which could be harmful. For those not accustomed to ice swimming, entering the water often triggers a reaction that makes one quickly get out of water. Breathing stops abruptly.

However, it's worth waiting a moment because after the first 30 seconds, the sensation changes. You notice that you can breathe and the condition does not worsen. When you bring your attention inward, to what you feel is warm, your state does not worsen, and you can still stay in the water. Calm breathing also helps. Of course, moderation must be remembered in cold water swimming. It's not a sport about competing over who can stay in the water the longest. That's why I like to swim in cold water alone. However, for safety reasons, it's recommended that beginners, at least, should not go into cold water alone.

I started ice swimming in the spring of 2020, when the world closed around us due to COVID-19. I needed an escape, over which only I could decide and feel the freedom to do so. Ice swimming is exactly this. I can freely decide to go into cold water and feel its power and energy. In the cold water, I reset, ground myself, and am present in the moment. I utilize breathing, meditation, and Reiki teachings.

THE BENEFITS OF COLD WATER SWIMMING

Cold water swimming – sometimes also referred to as winter swimming or ice swimming or simply cold exposure – generally means a quick dip in natural waters when the water temperature is under 50 degrees Fahrenheit (10 °C). Cold water swimming is known to be beneficial, and its popularity has surged in recent years. I started my own ice swimming hobby independently, but if one feels the need for help or support in getting started, there are courses and specific methods available for practice, such as the Wim Hof Method.

Dutch extreme athlete Wim Hof, nicknamed "The Iceman" after breaking many records related to cold exposure. Just to name a few, he has sat in an icy bath for 112 minutes, swum under ice for 66 meters, and climbed Mount Kilimanjaro in just shorts. While cold exposure is not exactly a sport, with these feats, he has demonstrated the enormous impact of the mind. Wim Hof has also been scientifically studied, leading to the conclusion that willpower can indeed influence the autonomic nervous system and the immune response. If you're interested in learning about experiences with cold swimming through a Wim Hof Method course, I recommend his book.

My favorite is also Finnish author Sanna Vaara's book *Mökkihulluutta ja mielenrauhaa*, translated in English as *Cottage Madness and Peace of Mind*. The book offers good tips on how to approach the water element and is written in a cheerful style, typical for Sanna Vaara. And, of course, it relates to the earth element as well, since the book talks about cottage life, a traditional Finnish way of spending summer holidays by the lake, often without electricity or indoor water, let alone indoor toilet.

We learn and grow into many of our perceptions, including our experience of cold. For example, in certain cases where a child has been raised by animals in the wild, the experience of cold of these children is completely different from that of children who have lived their entire lives clothed.

Wolf Siblings

Amala and Kamala were 8 and 1.5 years old, respectively, when they were brought to an orphanage in Bengal, India, in 1920. They had been found in a jungle cave living with wolves. It appeared they had lived with the wolves for a long time, as neither of the girls spoke and they behaved like wolves in other ways too. At night, they howled just like wolves and moved fastest on all fours. They ate only raw meat, scratched and bit when attempts were made to clothe them. In fact, they did not need clothes at all, as the cold did not bother them in any way. (L. Märcz)

According to studies, cold water swimming can have the following effects:

- Reduced experience of stress
- Improved sleep quality
- Increased mental clarity
- Enhanced immunity: significant decrease in colds
- Relief in experiences of pain
- Reduction in headaches and migraines
- Stimulation of blood circulation
- Better functioning joints
- Alleviation of allergy symptoms
- Skin firming
- Faster muscle recovery from exercise

Personally, I can vouch for almost all of these. In the last two years, I have started to run longer distances more than before. Previously, my problem was that my knees would become too stressed on long runs and hurt after. If I go into cold water after a run, there is no knee pain or muscle soreness at all. The body recovers better from the run.

There are also excellent books written on cold water swimming that offer great advice for beginners and those who have been swimming longer. To mention some of my favorites: James H. Smart's *The Cold Therapy Code: Rediscover Your Vitality Through Cold Exposure – The 3 Simple Cryotherapy Methods for Reducing Stress, Improving Sleep, and Increasing Energy* and Wim Hof's *Wim Hof Method book, Activate Your Full Human Potential.*

ABOUT SAUNA

I am not a true Finn, if I don't mention SAUNA in this book. The word sauna is a Finnish word. In Finland, there are over 3 million saunas, which is quite a significant number considering that the population of Finland is about 5.5 million people. About 90% of Finns go to sauna at least once a week.

Visiting cold water is often linked with the sauna – from one extreme to another. Some see the sauna and cold swimming as an absolute pair, while others would never venture into an ice hole after the sauna. I myself swim in cold water from the sauna when possible, but even prefer to do so without. However, it's generally not recommended to go straight from cold water into a hot sauna. The body should be given a moment to adjust to the temperature change. Initially, the body may experience stress from entering cold water.

Water is undeniably connected to the sauna experience. It is actually the water thrown on the stove that makes a sauna truly a sauna. There are many rituals associated with sauna bathing and throwing water on the stove. As a child, I learned to always throw the first scoop of water on the stove before entering the sauna properly. This steam is meant for the sauna elf, done so that the elf would be in a good mood and provide good steam, *"löyly"* in Finnish. Of course, one should not shout or argue in the sauna, as the sauna elf doesn't like that either.

The first known mention of a Finnish sauna is apparently from around the year 1100 in the chronicle of Nestor. The text is attributed to either the monk Nestor of the Kiev monastery or hegumen Silvester. The chronicle mentions "the hot saunas of the people living north of Novgorod and the custom of naked bathers to beat themselves with branches and finally pour cold water over themselves." This may sound weird but this (gentle) beating is still being done, especially when there is some special occasion, like mid-summer fest or Christmas sauna. Bunch of fresh, flexible branches of birch tree are tied together. They are called "*vihta*" or "*vasta,*" depending on which part of Finland you live in, and are used to scare evil spirits away.

This ritual resembles the sweat lodges of Native Americans. In some cultures, sweat lodges were used in purification ceremonies.

According to one estimate, there are 10 million saunas in the world, thus, one third of them are in Finland! Particularly in the first half of the 20th century, many Finns were even born in saunas. Previously, it was not customary to go to a hospital for childbirth. The sauna was just the right place for giving birth.

In a Finnish sauna, the air temperature can reach up to 212 °F (100 °C). The difference in the heat capacities of air and water is evident in the sauna; air at 100 degrees does not burn during a relatively short sauna session, but water at 100 degrees certainly does burn! As the air humidity increases when water is thrown on the stove to create steam, the specific heat of the air also increases. Although the sauna-goer sweats more, the temperature of the sauna does not increase due to the steam — "*löyly*"; it may even decrease due to the large amount of energy required for evaporation and vaporization.

WATER AND GROUNDING

I asked my daughter what water means to her. The answer was clear: above all, water signifies peace. I have come to the same conclusion myself. For example, if my thoughts are tangled after a long period of work, just stepping into the shower for a moment is enough. Suddenly, everything becomes clear, and ideas start to flow at the same pace as the shower. The mind calms down. I am able to reach the core of everything and my inner voice again. Water grounds us. It feels so natural that one doesn't even consider it a miracle. I wrote these lines here also as a reminder for myself, since these features of water are often easy to forget.

I highly recommend cold water swimming in seawater. All water is somewhat conductive, but seawater especially so. Conductivity is important for grounding. Most commonly, grounding is an electrical term that means connecting an electrical device or some other object to the ground with a conductor. Unwanted current is then safely directed into the ground, as in the case of lightning rods.

Grounding can refer to a spiritual grounding, often meaning calming down or forming a connection with one's inner self and simultaneously with the environment. It's also a phenomenon related to the material world – the body needs a connection to the ground, needs grounding. Grounding means touching the Earth. It can be walking barefoot outside, swimming, or even gardening, where you truly put your hands in the soil. These natural ways of grounding are of course the best. Products have also been developed for grounding that can have the same effect indoors, while sleeping, or even while working. I have used grounding sandals for years, as it's not always possible to walk barefoot everywhere. A grounding sheet and pillowcase can be tried for their effects on sleep. A grounding mouse pad is a better option than an ungrounded one, helping to relieve tension in the "mouse hand." Currently, as I sit here writing, I'm wearing a grounding bracelet on one wrist. A wire from the bracelet leads to a socket, through which grounding practically occurs, that is, through the socket's grounding. An electrical network is not used in grounding; the object to be grounded just needs to be connected to the ground with a wire.

In nature, grounding is related to moisture levels. Dry or painted concrete or rock does not ground well, nor does asphalt, which is a mix of refined bitumen from oil and aggregates. Dry wood, and of course plastic, do not ground either. Therefore, the best way to ground is indeed by going into seawater. Salty water contains minerals and electrons and is conductive – even hundreds of times more conductive than fresh water. Conductivity depends on the concentration of minerals in the water. However, all water is somewhat conductive.

When you stand outside barefoot, your feet interact with the ground. If there is a deficiency of electrons in the body, it seeks to balance and thus draws electrons from its surroundings, so, from the ground. In fact, such a deficit state can easily develop in modern times. The body is in a constant state of electron deficiency in chronic inflammation. Electrons are continuously needed to neutralize the effect of free radicals. Free radicals are atoms or molecules that also have an electron deficiency. Grounding actually makes the body more alkaline. If there are too many free radicals, they maintain inflammation. We are constantly exposed to free radicals. They are in food, air, and elsewhere in the environment. If you injure yourself, the number of free radicals increases. Free radicals increase inflammation, pain, and also all the effects of aging. Grounding is actually charging and caring for the body. In various inflammatory, pain, and other illness situations, it's worth ensuring adequate grounding, as it certainly does no harm.

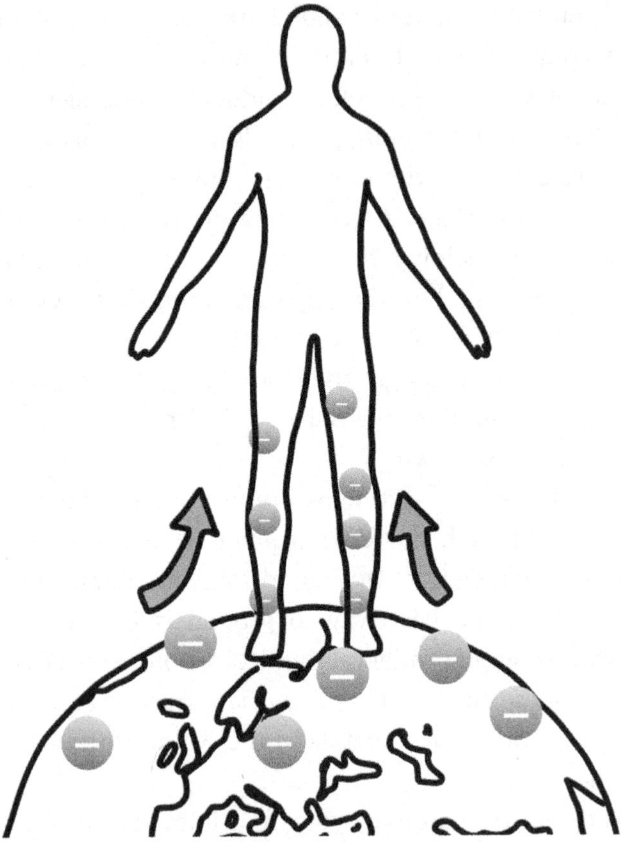

Figure 7. The body takes the missing electrons from the ground.

More on this topic can be read in the pioneering work on grounding by Clint Ober and others, *Earthing – The most important health discovery ever!*. The book includes many studies and case examples of the effects of grounding. Finnish author, Ph.D. Henna Helmenmaa writes about forest grounding in her book *Metsän hoitava syli*, in English *The Healing Embrace of the Forest*. Grounding in the forest is easier in summer, but in winter one can hug trees or try walking or running in wool socks. Dry wool is an insulator, but wool naturally tends to absorb moisture from the air among other sources. The wetter the sock, the better it also conducts electricity. However, running in wet socks might not be the most pleasant experience. It's best to find the grounding and relaxation methods that suit you. I, of course, also enjoy lying in a warm hot tub under the dark starry sky, but its mood-lifting effect seems to be based on something other than the effect of cold water. A plastic or wooden hot tub is not the best for grounding purposes, as it isolates from the ground.

~

Water is an essential part of grounding. Perhaps it's worth considering how water changes as it interacts with other materials or surfaces. Maybe the interaction causes it to change and react?

CHAPTER 5. ON THE EDGE OF THE WATER

"I want to re-emphasize that we're not dealing
with water at the molecular level;
we're dealing with crowds of water molecules.
We don't yet understand water molecules."

– Gerald H. Pollack, The Fourth Phase of Water

~

Water is a very sensitive liquid. Its behavior and properties seem to be significantly influenced by its environment and the types of materials it comes into contact with. And this includes all materials water touches – all surfaces, the container it's stored in, molecules dissolved in it, and even those it repels. Moreover, the effect seems to extend over a very long distance, thousands or even millions of molecular layers away.

Indeed, J.C. Henniker, in his summary from 1949, compiles over 100 studies that note the long-range interaction of water with various surfaces. Gerald Pollack brought this issue back into focus at the turn of this millennium. He also named the differently behaving zone observed at the water interface the EZ zone, or Exclusion Zone, based on how water behaves in it. His view is that this is a completely different type of water from what we are familiar with, and it could even be considered water's fourth state of matter.

Water behaves differently depending on whether it is in contact with hydrophobic or hydrophilic surfaces, that is, surfaces that repel or attract water, respectively. In the case of a hydrophilic surface, a thin layer, about the thickness of a hair (10 μm) or actually up to millions of atoms thick, of negatively charged particles forms between the surface and the water (Figure 8).

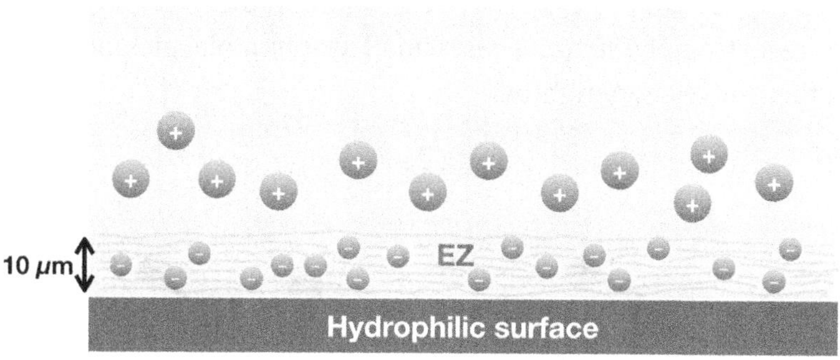

Figure 8. Water forms an exclusion zone near hydrophilic surfaces.

If there are other compounds dissolved in water, these compounds are unable to pass into this layer. The layer is quite thin, yet on the scale of molecules, it is truly vast. All hydrophilic surfaces create such a repulsion zone in water. Gerald Pollack and his team have studied the structure of surface water in more detail. In fact, it turns out it is not H_2O but, according to Pollack, a completely different kind of water with a honeycomb-like H_3O_2 structure (Figure 9). Ordinary water arranges itself into this form in the vicinity of hydrophilic surfaces. The structure resembles the structure of ice, but the layers formed from honeycombs are not attached to each other as in ice but can move freely relative to each other.

Among other materials, glass is a hydrophilic surface, as are paper, cloth garments, cotton, sugar, salt, limestone, and many other types of stone. Glucose has both a hydrophilic and a hydrophobic part, as does DNA, or more precisely, its amino acids. Hydrophobic materials include plastics, oils, fats, the leaves of many plants, bird feathers, vitamins, and cholesterol.

THE FOURTH PHASE OF WATER?

Professor Gerald Pollack from the University of Washington is one of the most well-known water researchers of the early 21st century. Pollack's research demonstrates how little we actually know about water and its interactions. He studied muscle function and wondered why prevailing theories explaining muscle function do not consider water, even though 2/3 of muscles are water. In fact, up to 99% of the molecules in the human body (about 70% by volume or mass) are water molecules, because water molecules are extremely small. These small water molecules are capable of organizing themselves very tightly.

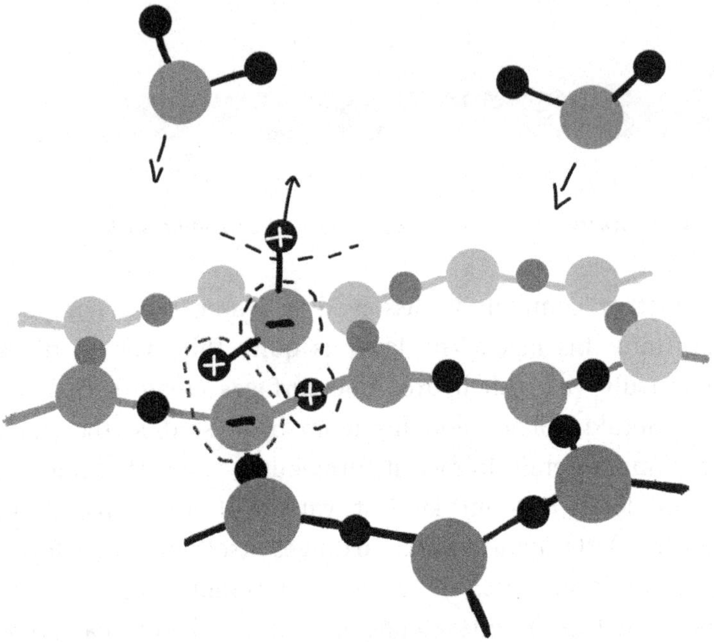

Figure 9. The formation of the honeycomb-like H_3O_2 structure. As the H_2O molecule arranges itself into this structure an extra H detaches.

In addition to muscle function, Pollack was also interested in cell function: why, for example, do proteins not leak out of the cell when the cell membrane is punctured? Or how can a cell remain alive even in freezing temperatures? After contemplating the subject and reading several forgotten studies in cell biology, Pollack concluded that the water inside cells must be different. It must be different from the water outside the cell. Inside the cell, water is more viscous and gel-like, and it repels molecules that do not belong there towards the cell wall, from where they are then expelled. The function of the cell wall could also be explained in this way. However, Pollack's hypothesis was in conflict with the current understanding of how the cell wall and cellular energy exchange work. After several years of further research, more and more interesting features of water have been discovered. He proposes that water has four fundamental properties that were not previously understood:

1. Water has four states of matter: ice, EZ (Exclusion Zone), water, and vapor.

Water behaves differently depending on whether it is in contact with hydrophobic or hydrophilic surfaces. A negatively charged EZ layer, millions of atoms thick, forms between a hydrophilic surface and water. Other compounds cannot penetrate this layer. This zone repels all other molecules, as shown in Figure 8. The water in this zone consists of interconnected H_3O_2 hexagons. This water is also known as hexagonal water, structured water, gel water or vortexed water. Its structure is like ice built from hexagonal cells, though it is liquid. From this structure, one could immediately deduce that such a liquid is more rigid than ordinary water, as the hexagonal structures do not "flex" and move in the same way. EZ water is thus more viscous. An

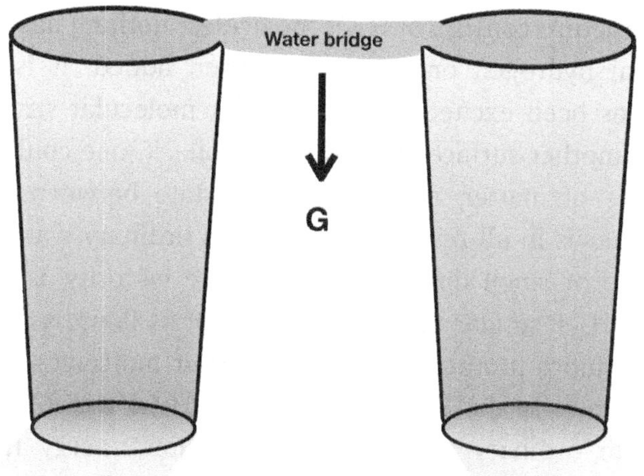

Figure 10. Gravity is not able to bend the rigid water bridge sufficiently.

example of EZ water's behavior and rigidity is a water bridge, where a bridge made of water forms between two glasses filled with water, connecting them (Figure 10). The water bridge remains intact as long as the stiffness of the EZ water is greater than the effect of gravity. A water bridge does not form between all glasses of water, but only when the water is EZ water.

The structure of EZ water is thus actually similar to ice: it is organized in the same way. Organized water seems to be present in living organisms: Mae-Wan Ho in her book *The Rainbow and the Worm*, writes about her studies on worms. She studied worms using a highly sensitive polarizing microscope, which can detect the organization of minerals, and found extensive structural organization in worms. Her conclusion was that this organization must be a result of organized water. Inspired by Ho's research, Pollack and his team studied EZ water using a polarizing microscope and found that the organized region is enormous relative to the dimensions of the molecules, amounting to as many as a million water molecules.

EZ appears to be an intermediate state that is also observed in nature when water freezes! Perhaps this explains why cold water feels so special? Ice does not form directly from ordinary water, but first organizes into EZ water, after which it changes its state to ice. Similarly, when ice melts, EZ water is formed first, which then turns into ordinary water.

EZ water is not chemically different from ordinary water, H_2O, even though its structure consists of H_3O_2 molecules; nothing has been added to it. No additional hydrogen or oxygen has been added; it is simply ordinary water that has been excited into a different molecular structure due to the influence of another surface. According to Pollack, one could even talk about a fourth state of matter, an intermediate state between ice and ordinary water. EZ water is in all respects the same as ordinary water. For example, it does not taste or smell different. It has more viscosity, is denser and more alkaline than H_2O, similar to ice, but ice is not as dense as EZ water, since in ice, there are more protons between the water molecules preventing tighter packing. All water in the human body is this H_3O_2 water. The body converts EZ water into ordinary water and also obtains energy from this process (Pollack, *Cells, Gels and the Engines of Life*).

Does water's surface tension come from EZ water?

Especially in calm weather, water striders – long-legged insects whose feet act like suction cups – can be seen on the surface of water, barely sinking into it and appearing to move on a film above the water. It is said that the exceptionally high surface tension of water makes it possible for water striders to stay on the surface.

Surface tension refers to the tension created by a single layer of molecules on the water's surface, which allows an object placed on the surface to float. In a liquid, molecules attract each other, and as a consequence to this attraction, the liquid stays intact without evaporating, meaning without molecules on the surface escaping into the air as vapor. On the surface, from the liquid's side, molecules experience a net attraction downwards, because there are no molecules above the surface to exert an attraction. This force tries to pull molecules away from the surface, and the surface area of the liquid tends to decrease because molecules are averse to be on the surface, resulting in tension on the surface. Since molecules interact more strongly on the surface of water due to hydrogen bonding, water's surface tension is also higher.

One consequence of surface tension is that small water droplets tend to take a spherical shape, where the surface area of the droplet is minimized. Surface tension can be reduced by adding substances like soap to the water.

The tendency of liquids to rise in thin tubes, or capillaries, known as the capillary action, is also explained by surface tension. The thinner the tube, the stronger the tendency to rise. The rise stops only when the hydrostatic pressure, the effect of the weight of the liquid column, becomes greater than the capillary action. For the same reason, water rises towards the top in the thin veins of a plant.

Studies have shown that water exists in two different forms: a low-density form with an organized tetrahedral structure, and a high-density form that is less organized. The ratio of these forms continuously changes due to various factors, such as substances dissolved in the water and temperature, and is also affected by nearby surfaces. Water behaves differently depending on what it is in contact with.

According to Gerald Pollack, the honeycomb-like organized structure that forms on the water's surface is very rigid. This structure could also explain how the Central American Basilisk lizard, also known as the Jesus Christ lizard, is able to walk on water. These lizards weigh about 7–14 ounces (200–400 grams) and can grow up to 2.5 feet (75 cm) long, including their tail. They can run on the water's surface at nearly the same speed as on land, about 15 miles per hour (24 km/h). Younger lizards can run up to 65 feet (20 m) on the water's surface before sinking, while older and heavier males run only 15 feet (a few meters).

The ability of the Jesus Christ lizard to stay on the surface cannot be explained by surface tension alone, as a single layer of molecules is not enough to support its weight. A layer thousands of molecules thick is required to keep the lizard afloat even temporarily. Gerard Pollack suggests that the Jesus Christ lizard stays on the surface of water due to its EZ structure. S. Hsieh and G. Lauder studied the lizard's running on the water's surface and found that the lizard's movement and posture play a role in its ability to stay afloat. The lizard's ability to stay on the surface is likely a result of all these factors combined.

2. Water gains energy from light.

Water is essential for us; without it, we generally cannot survive. There are exceptions that challenge this notion, such as the case of the Indian yogi Prahlad Jani.

The Yogi Who Lives on Light

Prahlad Jani was a yogi who claimed not to have eaten or drunk anything, including water, for decades. A group of doctors studied him in a hospital in Ahmedabad in 2010 for two weeks, trying to determine how Jani stays alive. At the time of the study, Jani was 81 years old. He spent the duration of the studies in his own room, under strict surveillance with cameras monitoring him around the clock. He did not eat, drink, or use the bathroom. His only contact with water was gargling and bathing periodically. At the end of the studies, the doctors could not refute Jani's claim. "We still do not know how he survives," neurologist Sudhir Shah told reporters at the time, "It remains a mystery what kind of phenomenon we are dealing with." The results of the study, funded by India's Defence Research and Development Organisation, were never sent to or published in a peer-reviewed publication.

Perhaps in Prahlad Jani's case, it's a matter of him gaining energy for his body in forms other than food or water? This is, of course, a completely revolutionary thought; maybe he is able to directly take energy into his cells from the sun? Pollack's research supports this idea.

Pollack suggests that the sun is capable of increasing the amount of EZ water in our bodies, thereby providing us with energy. He considers water so important to us that he presents Einstein's famous equation in a new form, $E = H_2O$, suggesting we should view water from this standpoint, as a source of energy.

Pollack's team observed that the EZ zone expands when illuminated. Ultraviolet light (UV), visible light, and infrared light (IR) all increased the EZ zone. IR radiation was particularly effective: after five minutes of IR radiation (at a wavelength of 3.1 μm, which corresponds to stretching of the OH bond), the thickness of the EZ zone increased more than threefold!

The EZ zone behaves differently than ordinary water because it reacts to light. The Sun's electromagnetic radiation increases the potential energy of EZ water. EZ water, therefore, gains energy from the Sun. Photons, or light particles, charge EZ water, increasing the order within and charging the water. Photons seem to make ordinary water molecules organize into EZ water, where the exclusion zone molecules are negatively charged, and ordinary water molecules are positive. This phenomenon seems to occur at the interface between water and another substance.

Water thus not only can receive energy but also use this energy! Indeed, when considered more carefully, this doesn't sound so impossible after all, since this is what plants do. Plants receive, absorb radiant energy, and use it. Since plants are mostly water, it no longer seems so miraculous that the same phenomenon occurs in a glass of water and the plant next to it. The water in the glass also transforms light energy into its own energy, just like plants. If you sit next to these, the natural conclusion might be that you also receive energy. After all, we humans are largely made of water. It could be assumed that our cells, which are EZ water, receive energy from the sun.

Water gains energy from light in the form of electromagnetic radiation, not only at wavelengths where IR radiation is most strongly absorbed. Perhaps this also confirms the benefits of the sauna and is part of the reason why spending time in an infrared sauna is good for health. Sunlight is more important to us than we have guessed.

Sound also appears to affect the growth of the EZ zone. Pollack's team exposed water to ultrasound and noticed that ultrasound at a frequency of 7.5 MHz initially narrowed the EZ zone, but immediately after the ultrasound was turned off, the EZ zone expanded several times, even five to six times more than before exposure.

J.W. Lee and G. Pollack have also studied the effects of wireless networks (Wi-Fi) on EZ water. The size of the EZ zone was measured during Wi-Fi exposure at various distances from the Wi-Fi source. The study found that the EZ zone reduced by 15–20%. The closer the water sample was to the source, the more negative the effect. Lee and Pollack state that, based on other studies as well, it is probably no surprise that Wi-Fi exposure likely has a negative impact on cell function and health.

~

3. Water is capable of storing energy.

EZ water is also able to store energy through its charge and organized structure. This could be an interesting approach to considering the possible memory of water. If water can store energy, perhaps it can also "remember"? EZ is negatively charged and can both store energy and release it as needed, like a battery or accumulator. It might, therefore, be possible to recover and convert light energy absorbed by water into electrical energy. With water and sunlight available everywhere, we would have access to freely available energy. In fact, water and energy cannot be separated from each other, so really, $E = H_2O$!

A "water battery," therefore, is charged by light. Just as plants convert the radiant energy they receive from the sun's photons into chemical energy through photosynthesis, so too does water receive energy. If this process could be made as efficient as photosynthesis, then harvesting the light energy of water could have a promising future. And this would indeed be genuinely free energy that we could use to solve energy shortages all over the globe, wherever there is water.

The formation of the EZ zone could also be used for water purification, as EZ water repels all impurities, thus purifying contaminated water. Collecting EZ water is therefore collecting clean water. According to Pollack, EZ water also repels salts, suggesting it could be used to purify seawater into drinking water. The process seems simpler than reverse osmosis – the process commonly used today for water purification and desalination. In reverse osmosis, water is purified by forcing clean water through a semi-permeable membrane from a more concentrated solution to a less concentrated one, the opposite direction to osmosis.

Were the Egyptian pyramids ancient power plants?

Have you ever wondered what the Egyptian pyramids really were? Often described as the tombs of the pharaohs, they are unlike other Egyptian tombs, like that of Tutankhamun, in that no mummies, sarcophagi, or other items typically taken to graves have ever been found in the pyramids. The pyramids are actually very bare, with hardly any hieroglyphs, which you wouldn't expect in a pharaoh's tomb. As a teenager, I was very interested in Egypt. I read everything related to ancient Egypt that I could get my hands on, starting with Mika Waltari's *The Egyptian*, and pondered the mystery of the pyramids. They include many fascinating aspects. For example, the three pyramids of Giza precisely mimic the arrangement of the three stars of Orion's Belt, The Three Kings. All three pyramids are also aligned with the cardinal directions with an accuracy difficult to achieve even with today's methods.

According to some hypotheses, the pyramids might have been power plants, perhaps based on hydraulics, in which fluids and pressure are used for power transmission. It has been suggested, for example, that the pyramids were hydraulic pumps into which water from the Nile was directed. This is supported by the existence of kilometers-long water tunnels beneath the pyramids connecting them and the ancient civilization.

This is all speculative, of course, and I am not an Egyptologist, but perhaps some such explanation could be more likely than the prevailing view that the pharaohs built such massive tomb monuments to bolster their own egos. In addition to water, the pyramids are often associated with another theme discussed in this book – spiritual growth. The pyramids have also been proposed as initiation sites for ancient mystery schools. By facing one's fears and looking death in the eye, remembering one's origin and end, a new path opens, and the traveler can step into the light. It is, of course, even less likely than the previous options that the pyramids were built solely as initiation sites.

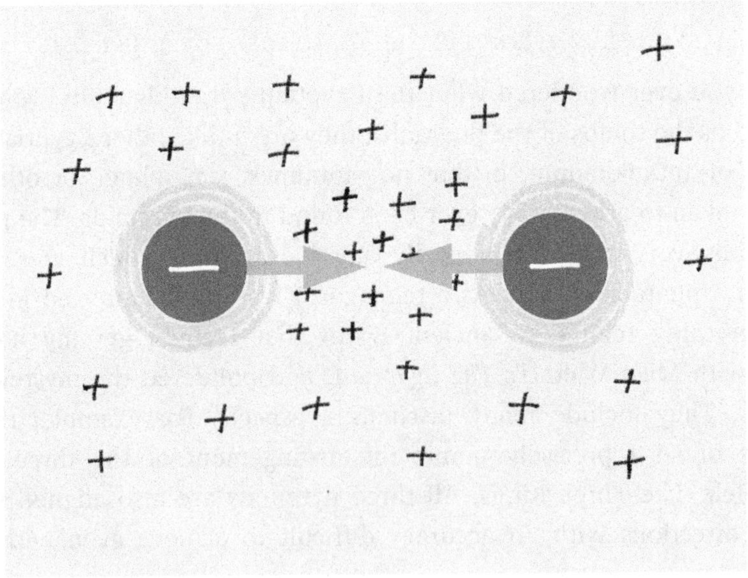

Figure 11. When an opposite charge accumulates between similarly charged molecules, they attract each other.

4. Similarly charged entities can attract each other.

Water molecules can form groups among themselves – water clusters. Two or more molecules with the same charge can join together. This might need some explanation, as it may seem at first that some laws of physics are being violated: when two similarly charged molecules are close to each other, an opposite charge begins to accumulate between the molecules (Figure 11). This causes the similarly charged molecules not to repel each other, but instead to start attracting each other and eventually unite. Thus, this does not violate Coulomb's law, which describes the behavior of electrical charges.

Physicist and Nobel Laureate Richard Feynman described this phenomenon with the phrase *"like-likes-like through an intermediate of unlikes."* Feynman's view was that such an attraction could be fundamentally important in physics and chemistry. One example is the formation of clouds. Clouds form from charged aerosol droplets. Traditional thinking would suggest that such droplets should disperse and spread, repelling each other, but according to the model described above, it is clear why the droplets attract each other and form a larger entity – clouds. The sun provides energy to the droplets.

The Law of Attraction

You may be thinking of the Law of Attraction which also mentions how like attracts like. The Law of Attraction is not a physical law, even though it mentions force, but rather an experiential, metaphysical law. It refers to how you attract exactly what you emit. What you focus on or concentrate on in your life, you also call towards you. Or, as Dr. Wayne W. Dyer stated: *"The law of attraction is this: You don't attract what you want. You attract what you are."*

Essentially, it means that what you think is possible and where you focus your thoughts and emotions also becomes possible. There are many sayings related to this, such as: "Ask, and it shall be given to you"; "The forest answers in the same way one shouts at it"; or "What you focus your attention on, strengthens."

Although the Law of Attraction is not directly discussed in physics, there is a basis for such ideas to be found there. We are not separate from reality but part of the whole, which we also affect. Many experiments in quantum physics have shown that the observer matters. The outcome of the experiment changes due to the observer's influence. This is exactly what the Law of Attraction depends on. We should learn to listen to our inner speech and think positively about ourselves and others. In *Hyperreality*, I wrote: "We are one, and everything you do, say, or think eventually comes back to you. ... You get what you give." Perhaps the Law of Attraction should be seen as a natural law?

HOW TO EXPLAIN EZ PHENOMENON?

The existence of the Exclusion Zone (EZ) has been experimentally confirmed in several laboratories, and now a theory explaining the phenomenon needs to be developed.

Professor Gerald Pollack's team has demonstrated that the EZ zone repels particles from the zone, such as micro-sized plastic particles and even fairly large molecules. The width of the EZ zone depends on the properties of the surface next to the water and environmental conditions and can be up to hundreds of micrometers. The interaction between water molecules seems to extend surprisingly far, even to millions of layers of water molecules, which nobody would have assumed. A single hydrogen bond lasts only a few billionths of a second, but new ones are constantly forming. The answers in the case of water are indeed not always found when looking at a smaller scale but when looking at the whole!

Physicist and materials science professor Rustum Roy, along with his colleagues, studied the melting of different materials and noticed that the surfaces of the materials had a certain organized structure when melted. According to them, this inevitably happens with water as well. G.N. Ling came to the same conclusion in his research: water molecules at the surface organize, and under ideal conditions, this orderliness can extend quite far. Among others, these studies support Pollack's observations.

Daniel C. Elton and others, in their publication "Exclusion Zone Phenomena in Water – A Critical Review of Experimental Findings and Theories," compile various studies that present alternative explanations to Pollack's hypothesis of water's fourth state.

Schurr's theory is based on the diffusion of colloidal particles and molecules in liquid. Colloidal particles are very fine solid particles in liquid, but present in such small amounts that the liquid appears homogeneous. It is not divided in any way. The theory predicts the growth of the EZ zone and changes in pH near a Nafion surface. The most common surface for generating the EZ zone in laboratory studies is Nafion, a sulfonated tetrafluoroethylene-based fluoropolymer with hydrophilic groups. Ordinary non-stick Teflon pans are made of polytetrafluoroethylene, by the way. Nafion contains plenty of protons ready to react with the liquid, leading to an inhomogeneous distribution of ions in the liquid. The liquid tries to restore balance, which results in particle movement away from the Nafion surface. Schurr suggests that OH^- or H^+ ions and salts, and changes in their concentration, are related to the formation of the EZ zone, a phenomenon known in relation to colloidal particles as diffusiophoresis. Huyghe and others propose that the EZ is created due to a combination of ion exchange and diffusiophoresis.

Recent studies continue to reveal even more interesting aspects of water. According to research by E.V. Del Giudice's team, EZ water can also act as an energy source (i.e., electron donor) for humans. Their research suggests that the body could convert EZ water into regular water, thereby releasing energy for the body to use, as Pollack also suggests in his hypothesis. It appears that Pollack's hypothesis, which suggests that a phase change or significant reorganization of water molecules into H_3O_2 water occurs in the EZ zone, does not fully explain all aspects of the phenomenon. The discussion on this phenomenon continues. In any case, the existence of the EZ zone has been confirmed by research.

Water indeed seems to undergo some special process that is not yet fully understood: water changes near hydrophilic surfaces, and it appears to have two different forms. Pollack's laboratory has been a pioneer in water research, and as noted in Elton's review article, other laboratories now need to conduct replication experiments, such as studying the growth of the EZ zone under laser illumination and how the zone repels salts. Neither of these phenomena is yet explained. Further research is also needed to understand water's behavior in Nafion tubes. Clarifying these mechanisms would surely help us understand similar phenomena in biology and find new applications. Water research continues, and most excitingly, many research groups have become interested in studying water in entirely new ways, inspired by Pollack's research.

Understanding water's behavior near surfaces may provide answers to many of water's anomalous properties. For example, the EZ zone and one of water's anomalous properties, the baffling Mpemba effect, seem to be connected.

The Mpemba effect

The Mpemba effect is the name of a phenomenon that has been known since the time of Aristotle: hot water freezes faster than cold water. Among others, René Descartes mentioned the phenomenon in his book *Discourse on Method* when writing about his vortex theory:

"One can see by experience that water that has been kept on a fire for a long time freezes faster than other, the reason being that those of its particles that are least able to stop bending evaporate while the water is being heated."

The phenomenon also seems to be related to supercooling, where a liquid cools below its freezing point but remains liquid. It has been proposed that the formation of hydrogen bonds might be related to these phenomena (Y. Tao, et al).

Today the phenomenon is named for the Tanzanian schoolboy Erasto Mpemba whose story became famous in 1963. He noticed the phenomenon while freezing ice cream during a home economics class. In a hurry, he put the hot mixture into the freezer, and the hot mixture froze before the cold mixture. Later, he asked a visiting lecturer, Dr. Denis Osborne, the reason for this and was laughed at by his classmates and teacher: *"If you put two identical containers with the same amount of water in the freezer, one at 35 degrees and the other at 100 degrees, the warmer one freezes first. Why?"*

Later, Osborne conducted an experiment confirming the same phenomenon. Osborne and Mpemba jointly published the experiment's results in *Physics Education*.

The Mpemba effect is challenging to observe, as its occurrence is influenced by many environmental factors and also, among other things, the freezer temperature, so don't be surprised if you can't reproduce the effect yourself. Several alternative explanations have been offered for the phenomenon. One proposes that the effect is related to dynamic equilibrium near the surface and the formation of water clusters in water, that is, the formation and behavior of the EZ zone. The phenomenon is actually easy to understand because the water molecule has a rigid H-O-H arrangement and strongly resists the formation of a crystalline structure. At the same time, water molecules cling more tightly to each other and to hydrophilic surfaces – those that tend to orient molecules. Thus, the EZ zone is formed.

A similar phenomenon also occurs in capillary tubes, described above in chapter 2. When water was heated and cooled in thin glass capillary tubes, it behaved in a completely new way, and the properties of water in capillary tubes change remarkably. Research by Pollack and others on EZ water actually also vindicated Derjaguin's reputation, as studies in capillary tubes were now proven correct.

CHAPTER 6. CAN WATER REMEMBER?

"Water has a memory and carries
within it our thoughts and prayers.
As you yourself are water,
no matter where you are,
your prayers will be carried
to the rest of the world."

– Masaru Emoto

~

On a chilly October day, I went swimming in the sea, silently asking for a sign that my book was heading in the right direction and would contain what is essential for us all. Writing books is for me a process where, at best, I feel I am merely a conduit for information to flow through. I have sought signs and synchronicities when writing my previous books as well. My surprise was considerable when, swimming back to shore, I encountered a jellyfish swimming beside me! A beautiful, pale pink creature with hints of lilac, about the size to hold in my palm. I was swimming in the coastal waters of Helsinki. Could one hope for a better sign?

HOW CAN A JELLYFISH REMEMBER?

Jellyfish are essentially water. At 98% water, they are like a transparent jello. Yet, some of them, such as the box jellyfish (Lat. *Cubozoa*), are known to possess at least some form of short-term memory and the ability to learn, even though they lack brains. In fact, they do not have hearts either. Yet, a jellyfish is a fully functioning living being. How is this possible?

Jellyfish are incredible creatures. I am certain that studying them and nature in general would provide answers to many questions that are still open. Indeed, research on jellyfish has already provided a wealth of information and new applications. Researching jellyfish has even led to a Nobel Prize: many jellyfish are known to be bioluminescent, meaning they can illuminate themselves. An important part of this process is GFP (Green Fluorescent Protein), for the discovery and development of which O. Shimomura, M. Chalfie, and R. Tsien received the Nobel Prize in Chemistry in 2008. The protein can be used, among other things, as a marker in the human body to study various bodily processes, such as insulin production or muscle structure.

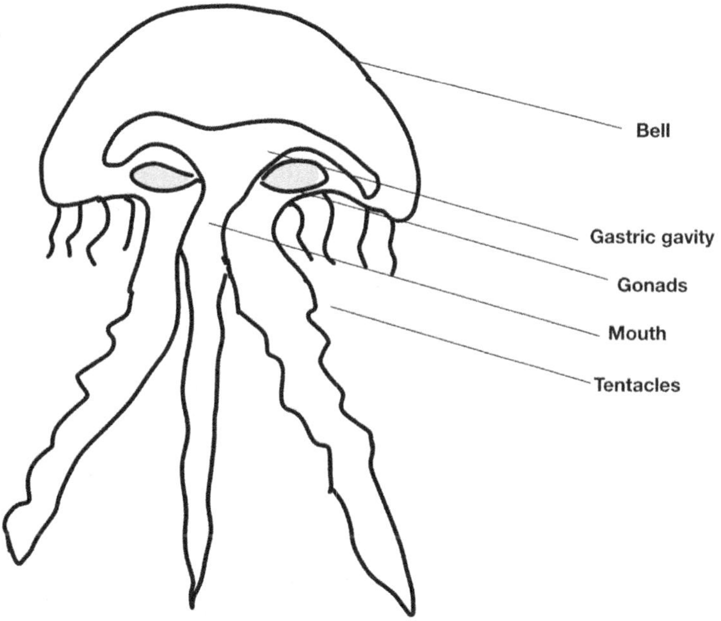

Figure 12. The structure of a jellyfish.

Jellyfish have been around for about 500 million years, meaning they have survived many upheavals. They were on Earth long before the dinosaurs. The jellyfish body, or "bell," which is the most visible part, consists of a gelatinous structure only a few thin layers of cells thick.

Many jellyfish have four nearly circular structures visible in the middle of the bell. These are the gonads, surrounded by four oral arms and catching tentacles with stinging cells. The venom of some jellyfish is very dangerous to humans, so dangerous that if you eat a squid that has eaten a jellyfish and has not yet digested it properly, you could get poisoned and even die. However, most jellyfish are harmless to humans, like the moon jellyfish (Lat. *Aurelia aurita*) I encountered on my swim.

Jellyfish can reproduce both sexually and asexually. Depending on the species, they can either create a copy of themselves by splitting or the more common way, eggs released into the water by females are fertilized by males' sperm, after which they begin to grow. Amazingly, there is a species of jellyfish that can rejuvenate itself! Molecular biologist Nicklas Brendborg from the University of Copenhagen describes in his book *Jellyfish Age Backwards: Nature's Secrets to Longevity* a jellyfish called the "Immortal jellyfish" (Lat. *Turritopsis dohrnii*), which can reverse the aging process. Instead of aging and dying, they can return to their youthful, polyp state. If a jellyfish becomes stressed, for example, due to hunger or an unexpected change in water temperature, it can revert from its adult form back to a young polyp form. Its aging process is reversed. No one knows how the jellyfish does this. The process is called transdifferentiation. The rejuvenated individual can then age again, and it can perform this rejuvenation repeatedly. Essentially, it's as if a butterfly could revert back to a caterpillar, or perhaps us back to babies!

It is thought that many processes, including aging, are unidirectional. For example, once a cell reaches a certain state, it is not possible to reverse its development. However, Japanese researcher Shinya Yamanaka has shown that specialized cells, which were thought to have reached their terminal state, can revert back to pluripotent stem cells, which can differentiate into any tissue cells. In 2006, Yamanaka identified a small number of genes in the mouse genome that were significant for the process. Activating these genes allowed mouse skin cells to be reverted to undeveloped stem cells. The process also rejuvenates the biological age of the cell. Yamanaka later received the Nobel Prize in Medicine in 2012 for this discovery. It is probably the same process by which the Immortal jellyfish rejuvenates itself.

We know which genes to activate, but not what actually triggers their activation. Nor can we explain how stem cells end up specializing into specific cells, such as cardiomyocytes, kidney cells, or neurons, among others. Where is the information that tells them where they need to specialize?

Indeed, jellyfish do not have brains. They sleep, even though sleep is thought to be related to memory and the brain's REM cycles. They do not have hearts, but this is understandable since they do not have a circulatory system either, so a heart is not needed to pump blood. What is special about jellyfish is the complex network of neurons that covers their entire body. The neurons of the jellyfish have various tasks. There is a broad motor network that enables it to move and a sensory network known as rhopalia that allows it to sense and react to stimuli. Its neurons extend everywhere, not just on the surface but also into the inner organs of the jellyfish. With the help of neurons, the jellyfish can sense light, touch, temperature, gravity, and even the salt content of water and react accordingly. Research on jellyfish has concluded that the neuron network must also serve as a substitute for centralized brains. However, it remains a mystery how information is transmitted in different parts of the jellyfish or from the neurons to other parts of the body. Information is generally assumed to be transmitted through the brain, but since jellyfish do not have one, it's a mystery how they can, for example, seek shelter, mate, change the environment, and make other decisions that require some form of choice.

Why is the neuron network often considered an exhaustive explanation? Perhaps water, which constitutes 98% of a jellyfish, also plays a crucial role in storing and transmitting information? T. Morawietz explained the behavior of water and its far-reaching interactions using a model based on neural networks. Could understanding water explain the internal flow of information in a jellyfish and how information is retained?

In *Hyperreality*, I discussed the planarian flatworm, a species that has been on Earth for hundreds of thousands of years and can duplicate, or clone, itself. If the worm is cut in half, it grows two identical living individuals from the halves. The newly cloned individuals can "remember" their original habitat, among other things. Information thus also passes to the new individuals. When there is enough food, Planaria grows and ages in the usual way, but if there is no food, it starts to consume itself, beginning with the least essential parts, and stops when only the nervous system is left. This buys it more time, if conditions improve. When times are better, Planaria can regrow the missing parts and live normally again. In both the jellyfish and the planarian cases, the same question arises: where is all the information really located? Answers are rarely sought in something as common and ordinary as water.

WHAT DID BENVENISTE FIND?

Jacques Benveniste was the first to raise the concept of water memory. Benveniste's conclusion was that water is capable of retaining information. Water "remembers" which molecules it has interacted with.

I have already mentioned Benveniste's studies and their reception earlier in this book. Unfortunately, Benveniste's case is not unique; throughout history, it has been common for new ideas to be easily dismissed. In my previous books, I have discussed similar examples, including Ignaz Semmelweis, William Harvey, and perhaps the most famous example to many, Galileo Galilei. He was placed under house arrest for the rest of his life for not conforming to the Church's opinion. Semmelweis's conclusion about the connection between the dirty hands of students and doctors and childbed fever was dismissed – bacteria were not yet known about. Harvey also faced dismissal when he proposed that the heart acts as a pump circulating blood – the heart was just considered a heater. I have also mentioned a similar attitude in my book *From Quantum Physics to Energy Healing*, when discussing William Bengston's energy healing research on mice. His research is practically ignored because it cannot be explained.

It is sad that researchers still face such treatment when openly raising questions in a desire to understand better. If the subject or the framing of the question does not fit into the prevailing paradigm, it is most likely to be dismissed. Questions should be asked, and answers sought without being stigmatized. Why are views that challenge prevailing thought dismissed? Benveniste himself asked in his article in *Nature*: *"What if I am right?"*

What did Benveniste actually find?

Benveniste's article, published in *Nature* in 1988, dealt with the effects of highly diluted anti-immunoglobulin (anti-IgE) antibodies on the activity of human basophils. Basophils are white blood cells that become activated in allergic reactions when an allergen attaches to an IgE antibody on the surface of a basophil or so-called mast cell.

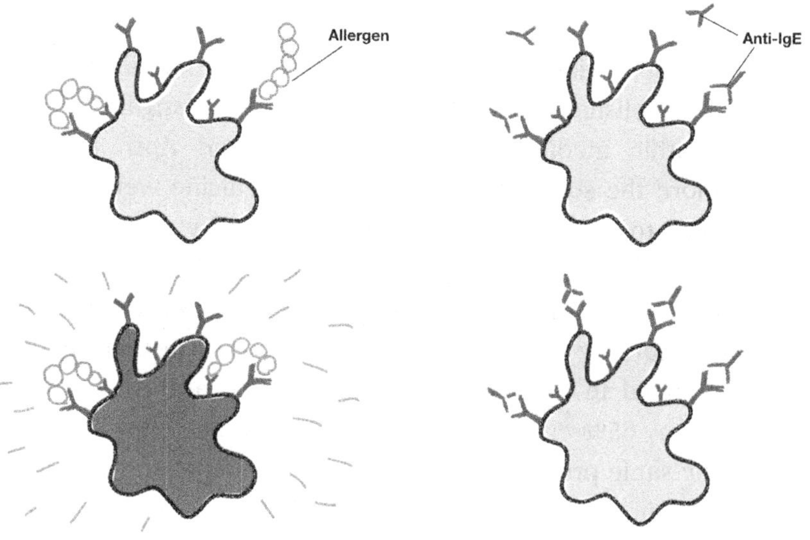

Figure 13. On the left (upper picture), an allergen attaches to a basophil, activating it, and (lower picture) histamine is released. On the right (upper and lower picture), the anti-IgE antibody inhibits this process.

As a result, histamine is released into the body, causing an allergic reaction. The anti-IgE antibody blocks this action, which, in the best case, means no allergic reaction occurs.

The dilutions were made from 1 x 10^2 up to a very high dilution of 1 x 10^{120}. In the highest dilutions, statistically, nothing of the original substance was left, only water molecules. However, in the highest dilution, a clear activation of white blood cells by 40–60% was observed, even though there were no IgE antibody molecules in the dilution! The dilution process involved vigorously shaking the dilutions. In the publication, Benveniste and his team ask whether shaking could help in transferring biological information to water as its molecules reorganize.

Homeopaths naturally welcomed Benveniste's studies, as the procedure described by Benveniste is similar to that used in homeopathy. In the treatment, the substance used is diluted to the point where it is practically no longer present in the final product. In homeopathy, this process is called potentization. Benveniste was not so much interested in homeopathy but in science.

In a scientific publication related to the field of pharmacy also from 1988, they nevertheless published their results from studies investigating the effects of two homeopathic medicines (*Lung histamine* and *Apis mellifica*) on allergies. The more the solutions containing the medicine were diluted, the better they seemed to work, which is completely contrary to what one would expect. Usually, such solutions, known to have a low molecular concentration which are insufficient for producing a biological effect, are not even used in studies. Perhaps learning from the case of polywater, Benveniste and his group also mentioned in their publication that air or a glass vessel do not affect the results, as 65% ethanol and the saline solution used in the studies went through the same process as the homeopathic medicines, and no traces of either were seen in the dilutions.

The diluted solutions, practically just water and its water molecules, seemed to be structurally active. Benveniste used the term "water memory" in his hypothesis. The same effect was achieved by pouring either the original solution or several times diluted solution onto the cells. And actually, the most diluted solutions worked even better than those that were only slightly diluted. The diluted solution seemed to retain a "memory" of the molecules with which it had interacted. The diluted solution, practically water, had the same effect as the molecules originally in it.

In response to the criticism received, Benveniste pondered in his article in *Nature* why the notion of water memory had received such a reception. Their group had done meticulous work and wrote the publication for two years. In that time, the same results obtained by five impartial laboratories in Canada, France, Italy, and Israel (two laboratories) were also dismissed and even seen as a fraud. Benveniste believed the answer lay in the discovery's connection to homeopathy: from water memory one might draw conclusions in support of homeopathy, which might be seen as threatening to established Western medicine.

Later, in 2000, a consortium formed by four independent research laboratories in France, Italy, Belgium, and the Netherlands repeated Benveniste's experiments again. The experiments were conducted as double-blind tests. The solutions under investigation were prepared in external laboratories, and the consortium research laboratories did not know whether the solutions added to the study's basophil-algE reaction contained only a "memory" of histamine or if they were pure water. The entire series of experiments was coordinated by an impartial researcher, who coded the solutions and recorded the data but did not otherwise participate in the study in any way. The test was entirely objective, so the truth could finally be revealed.

As a result, three out of these four laboratories reported that the results were statistically significant and confirmed that an extremely diluted solution previously containing histamine indeed produced the same result as a solution containing the substance! The results of the fourth laboratory were also very close to the threshold of a statistically significant result, meaning all four laboratories confirmed the results of previous experiments.

Professor Madeline Ennis from Queen's University Belfast, who participated in the research consortium, had been very skeptical of Benveniste's results from the beginning and joined the consortium wanting to finally prove definitively that Benveniste was wrong. It turned out differently. Ennis states in The Guardian's article about the experiments: *"Despite my reservations against the science of homoeopathy, the results compel me to suspend my disbelief and to start searching for a rational explanation for our findings."*

The Guardian also interviewed Benveniste, who, however, was not very impressed by the results. Even though it had been 12 years since his results were reported, the situation remained the same: the same experiment was still being repeated!

How then could water possibly store information? Specifically, Benveniste did not seek the reason for his findings in his publications but only stated that water could act as a kind of template, a model that enhances the antibody molecule through a network of hydrogen bonds or through electromagnetic fields.

The suggestion of hydrogen bonds playing a possible role in the transmission or storage of information is often dismissed because hydrogen bonds are very short-lived, lasting only on the order of picoseconds (10^{-12} seconds) before they break and form again. Therefore, the possibility that they or their combinations could somehow store information for a longer term, such as related to antibodies, is considered impossible. However, as hydrogen bonds continuously form anew as old ones break, could there be a kind of dynamic memory that continuously carries and transfers information? It is worth remembering that, in addition to hydrogen bonds, water also generates long-range interactions, which could also have an effect on information storage. And indeed, according to neuroscientist Rodolfo Llinas, we know almost everything about the brain, except how it works. Perhaps an understanding of water could also help in this area?

Benveniste continued his research with his team and published an article in the journal *Medical Hypotheses* about how molecular activity could be transferred electromagnetically through water. They digitally recorded (sample 44 kHz) biological activity on a computer. When this recording was played to water, cells, target organs, or even allergens after exposure, the reaction was similar to that of the original substance. In their experiments, even bacteria could be activated this way. Based on several hundred experiments conducted, they were convinced that it is possible to transfer specific molecular activity-enhancing information to water.

Allan Widom, Yogi Srivastava, and Vincenzo Valenzi suggest that biochemical information could be stored in water, in its electric dipole moments in the same way that magnetic moments store information on magnetic tape or a computer's hard drive. Electromagnetic signals could thus enable the transfer of molecular information between molecules.

What I have described sounds incredible. However, research shows that certain frequencies affect cell function and can therefore help restore or maintain health. Benveniste was the first to report the "digitization" of biological signals, and this has been confirmed in later studies. Luc Montagnier's group, for instance, demonstrated how certain bacterial DNA sequences have the ability to induce electromagnetic waves in highly diluted, i.e., homeopathic solutions. Alberto Foletti and his team showed how information travels between human cells electromagnetically in aqueous systems. Also interesting is a study led by Hans Jonas, which dealt with Benveniste's findings. The group consisted of both researchers skeptical of Benveniste's findings and those more open to them, including researchers from Benveniste's team. However, they were unable to replicate the results, except in an experiment conducted by a researcher from Benveniste's team.

No explanation was found for this. In fact, Benveniste himself noted in one publication that some researchers produce clear changes, some none at all, and some seem to block the effect. This is also systematic, happening time and time again. However, the group did not test Benveniste's suggestions for why this might be. They simply state that it's possible that some unknown factors could be observed in the results, such as the effect of chemical residues or the intention of individual experimenters or some kind of energetic effect.

How were biological signals then digitized? The Jonas group describes the digitization process in their publication "Can specific biological signals be digitized?" as follows: *"The sample to be recorded is placed between one transmitting and two receiving coils. The transmitting coil produces an electromagnetic noise signal that is passed through the sample and received by the receiving coils. These two receiving coils drive a differential amplifier whose output is connected to the 'mic' input of the computer sound card and the recording is achieved as is done with normal sound through a microphone with settings at 44.1 kHz, mono, for 3 s."*

The source of electromagnetic signals was thus a certain biological activity, during which recording was activated. The results of the experiments seem to support the idea that biochemical information can be transferred electromagnetically in solution.

MONTAGNIER AND THE ABILITY OF WATER TO TRANSFER INFORMATION

Later, among others, Professor Luc Montagnier, a researcher who received the Nobel Prize in Medicine in 2008 for the discovery of the HIV virus, has proposed the idea that water has the ability to carry information. Montagnier published two articles in 2009 about his observations on the peculiar properties of water. If these results are true, they are said to be the most significant experiments conducted in the last hundred years, leading to a completely new examination of modern chemistry. In his publications, Montagnier presents something truly revolutionary: highly diluted DNA is able to transmit information about itself as if "teleporting" itself to another test tube.

The discovery in Montagnier's laboratory occurred somewhat by accident. In one test tube, there was a highly diluted solution containing DNA from certain pathogens, and in another test tube, there was pure water. The test tubes were subjected to a low-frequency (7 Hz) electromagnetic field for 18 hours, after which it was noticed that the water in the tube initially containing pure water now also contained the same DNA information as the other tube! The information had been transferred from one test tube to another. Both now reacted similarly to the original DNA solutions.

In his 2011 publication "DNA Waves and Water," Montagnier approaches the topic through quantum field theory. According to him, water should be viewed not only in terms of static interactions such as hydrogen bonds and electric dipole-dipole interactions but also as an electromagnetic field. The phenomenon observed by Montagnier is evidence of quantum phenomena in water. At the quantum level, phenomena in water are visible on an everyday scale, and information can be transmitted over long distances! Water seems to be a good medium in which DNA can replicate itself and transmit information, signals, through quantum entanglement. Montagnier suggests that this could partially explain the efficacy of homeopathic remedies. According to him, it seems that as the original molecules are removed through dilution, the structure of water begins to mimic these molecules. This new information could be utilized in many ways; Montagnier mentions virus detection as one example.

The phenomenon presented by Montagnier appears to have similar characteristics to the entanglement phenomenon familiar from quantum physics, for which the Nobel Prize in Physics was awarded to three physicists, Alain Aspect, John F. Clauser, and Anton Zeilinger, in 2022. Entanglement is a phenomenon where two particles maintain a connection even when far apart, even kilometers away. For example, if the spin, characteristic for particles, is changed in one particle, it immediately changes in the other entangled particle. The phenomenon was first theoretically described by Schrödinger in the 1920s and experimentally confirmed in the 1970s and 1980s. Entanglement is evidence that non-local phenomena really exist and that information can travel instantly from one point to another. Entanglement was originally observed for individual particles, but the limit has not yet been found for how large a system in which it is possible. Particles act as a unified whole, like a flock of birds or a school of fish, so why not two solutions, or even water?

It is interesting that the signal in which the phenomenon is activated in Montagnier's studies is at a frequency of 7 Hz. That frequency is close to the fundamental frequency of the Schumann resonance for the Earth's electromagnetic field, which is about 7.8 Hz. The same frequency is also at the boundary between the Theta range (approximately 4–8 Hz) and the Alpha range (8–13 Hz) of brainwaves, that is, at the border between sleep and wakefulness. Hypnosis as well as emotions and imagination are experienced in the same range of brainwave frequencies, and the state is one of deep relaxation.

ABOUT HOMEOPATHY

Homeopathy, which Montagnier mentioned in his studies, operates on the principle of *"like cures like"* (Lat. *similia similibus curentur*), meaning that the same substance which causes or has caused symptoms of an ailment or disease can also cure them. In fact, modern medicine sometimes utilizes a similar principle: quinine, used to treat malaria, for instance, induces the same symptoms as malaria in a healthy individual!

Physician and chemist Samuel Hahnemann, considered the "founder" of homeopathy, stated in 1796 that diseases can be cured by potentizing substances, that is, gradually diluting them and vigorously shaking them at the same time. Homeopathy quickly gained popularity, and its effectiveness was documented. During the 1918 flu pandemic (Spanish flu), for example, in the United States, the mortality rate among those using homeopathy was only 1.05%, compared to over 30%, occasionally even up to 60%, mortality rates with Western medical methods. During World War I, the U.S. forces had their own medical units practicing homeopathy, with 100 nurses, 22 doctors, and a couple of dentists.

As Western medicine developed, the popularity of homeopathy declined, but in recent decades, its popularity has begun to rise again along with the interest in holistic and natural treatments. Especially in Central Europe, homeopathy is common, with pharmacies having dedicated sections for homeopathic remedies, and many doctors incorporating homeopathy into their practice. In Switzerland, homeopathy is highly regarded, and only doctors who have completed additional homeopathic training alongside their medical studies are allowed to practice homeopathy, as it is classified as a medical treatment in Switzerland.

In homeopathy, the memory of the causative agent of the disease is used to cure the illness. Homeopathic treatments do not directly medicate, as in conventional medicine, but rather, as in many other complementary treatment methods, they activate the body to heal itself. Such a process resembles vaccination, where a weakened virus or bacteria, or their components, are administered to induce an immune response. The difference, of course, is that in homeopathy, often the original substance is no longer present in the dilution. And this is precisely the point that challenges the linear cause-and-effect thinking of Western culture: if the active ingredient is absent, how could it have an effect, as surely just water cannot retain information?

However, could the process used in homeopathy – extreme dilution and vigorous shaking – affect the long-range interactions and structure of water, and thereby its properties?

In the preparation of homeopathic remedies, potentization is a crucial part. The original substance is diluted in the correct proportion with water and vigorously shaken. Something mysterious seems to happen when solutions are diluted to extreme degrees and vigorously shaken. Water plays a significant role here as well. Thus, finished homeopathic remedies may contain the original substance, but often the dilution is so great that they do not contain it at all. Claims that homeopathy does not work often relate to these very high dilutions. These are similar claims to those found, for example, in the studies of Benveniste.

Millions of people around the world benefit from homeopathy every day, and in Europe, one in three people uses homeopathy. Therefore, we should try to understand what homeopathy is really about. Sadly, in recent years, many national bodies, such as those in the UK, France, and Australia, have declared that there is no scientific evidence for homeopathy. They are, thus, considered ineffective, and these bodies recommend against funding research into homeopathy. But why? Wouldn't research be needed to understand what homeopathy is about? Currently, the challenge in homeopathy research, as with other complementary treatments, is the lack of funding and the difficulty in getting articles published on the subject in reputable scientific journals. It seems that the fundamental purpose of science, which is to question and understand phenomena, is being forgotten.

As citizens, we should share our experiences and opinions and encourage unbiased research. The purpose of research is to find answers to various questions that may even differ from the prevailing paradigm. We do not learn anything by only walking down already trodden paths; sometimes we must dare to look beyond the hedge or into the ditch, metaphorically speaking, to find something interesting. At the same time, it should be noted that researching homeopathy is challenging, and it is most likely not possible to study it using the same methods as pharmaceutical treatments. It may not be understood solely through a materialistic approach. Phenomena must be studied on their own terms.

There has been some scientific research of homeopathy. Studies have collected users' experiences of the effectiveness of homeopathy in treating many diseases and ailments. In the reference list, you will find a link to a list compiled by holistic and complementary medicine researcher, Dr. Iris Bell. She holds doctorates in medicine and philosophy from Stanford University and also licenses in homeopathic and integrative medicine. She has researched and used complementary treatments in her work for over 30 years. I highly recommend also watching the documentary "The Magic Pill" by filmmaker Ananda More, which explores homeopathy.

So what is the mechanism of action in homeopathy? There is no answer yet, but many projects are gradually approaching this question. For example, the physical and chemical properties of extremely diluted solutions are being studied to determine what process affects the final results. Another point of interest is the effect of vigorous shaking, as it seems to be an essential part of the process and affects the results.

The challenge is to explain how information is preserved, especially in solutions that have been diluted to extreme degrees, beyond the so-called Avogadro's limit, and how the information is transferred from water to living systems. One proposed hypothesis is the nanoparticle hypothesis, because extremely diluted solutions have been found in experiments to contain extremely small metal nanoparticles.

J.R. Pereira Guedes et al. consider different hypotheses for the transmission of information and present the entanglement phenomenon, familiar from quantum physics, as one alternative hypothesis. Once a substance has been present in a solution, the connection between the water and the substance is maintained even if there is no trace of it left in the diluted solution. Perhaps entanglement occurs here, as well as between the diluted solution, the original substance, and if it concerns homeopathy, also between the patient and the doctor? Everyone is connected and intertwined – perhaps mere connection is enough for essential information to be transmitted?

A. Manzanini and B. Galeazzi explain homeopathy through quantum electrodynamics (QED). The QED method describes the interaction between electrically charged particles and electromagnetic fields, such as how light and matter interact. They created a model based on QED theory for the homeopathic system they studied. In the model, the living organism is thought to consist of its characteristic wave functions, which encompass all possible levels of the organism, forming a multi-level coherent whole. When coherence is disrupted, the organism becomes ill. They propose a mechanism by which information could be transferred from homeopathic water to a multi-level coherent living organism, such as a human being.

Glen Rein also considers QED as a possible explanation for homeopathy. He has conducted UV spectroscopy measurements, studying water energized in various ways. Changes in water spectroscopy images support his hypothesis that quantum fields play a role in how information is preserved in water. What may also be surprising is that in his studies, the change in water appears to be permanent, lasting for months! One could assume that any potential change would be only momentary and that water would return to its previous state. However, this does not seem to be the case; water retains information about its history and, among other things, about the materials it has interacted with.

As Montagnier pointed out, something peculiar seems to occur in water when it is diluted to extremes. Water appears to begin mimicking the structure of the original molecules. Physicist Brian Josephson, also a Nobel laureate, reminds us that it is often thought that because water is a liquid, it cannot have the structure needed to explain these phenomena. But why couldn't it? There already exist liquid structures with specific organization: liquid crystals, used, for instance, in liquid crystal displays. They flow like liquids but still maintain organized structure over long distances. According to Josephson, water has the ability to "communicate" over distances, and such interaction can be described by quantum theory and electromagnetism. Perhaps the EZ zone and its organization could be an especially significant phenomenon for homeopathy?

~

Meteorologist Edward Lorenz stated in 1963 that small changes in initial conditions can amplify and make long-term prediction impossible. He also discovered a factor that amplified disturbance. The phenomenon is famously known as the butterfly effect, which he described as follows: *"The flap of a butterfly's wings in the Amazon can cause a hurricane in the USA."* Initially, Lorenz used the analogy of how a seagull could cause a storm at sea but changed it to the more famous form because it seemed to stick better in people's minds. Factors that cause such a butterfly effect can form fields, such as gravitational, magnetic, electric, or quantum fields. Fields are invisible but connect. All information and experiences are conveyed through fields. Rupert Sheldrake's morphic fields theory, among others, describes the behavior of such fields in a similar manner.

It is known that there is a connection between life and water – wherever there is water on Earth, life has been found. Therefore, it can be said that water is the medium through which life manifests itself. Perhaps the butterfly effect can be observed in homeopathy: if active molecules are removed from water, the information about them still remains in the water, and the most diluted becomes the strongest.

MASARU EMOTO'S ICE CRYSTALS

Right now, as I'm writing this, enormous snowflakes are falling from the sky. It's December, and there's already over half a meter of snow on the ground, which has been rare here in southern Finland in recent years. This year, it seems like Christmas will be white. As a child, when the first snowflakes fell, I remember watching them in wonder: each one is different, there are no two snowflakes alike. They are like people; there are no two alike. Somehow, it felt like there's a core to everything.

Water holds secrets that we have not yet learned to understand. How does it manage to create such beautiful snowflakes, and above all, why does it create them? In nature, nothing is pointless; everything has a reason. Does water reveal something more about itself through snowflakes, ice crystals, perhaps a code, a key to everything?

Masaru Emoto, an internationally renowned Japanese water researcher, conducted groundbreaking research for years related to consciousness and water. Emoto suggested that consciousness, intention, affects water and the ice crystals formed from it: ice crystals formed from water exposed to positive words and emotions are more symmetrical, harmonious, and simply more beautiful than those formed from water treated with negative words and emotions. According to Emoto's research, water has the ability to store information.

Emoto's studies have naturally intrigued many, but at the same time, they have also received much skeptical criticism. The idea that consciousness could influence the physical world feels foreign to many and does not fit into scientific materialistic thinking. Emoto's research has been criticized for insufficient experimental control data and for not documenting or reporting his experiments accurately enough. There have also been possibilities of manipulating results or making mistakes. In 2003, James Randi even offered Emoto the chance to earn a million dollars if he could replicate the experiment under test conditions agreed upon by both parties. Naturally, Emoto did not warm up to the proposal.

Many don't understand this. If you're doing something you know works and is true, why wouldn't you prove it to everyone else and at the same time earn a million? However, thoughts and emotions have immense power. Just the thought of conducting an experiment in conditions where at least some intend to prove what you're doing is a hoax, doesn't excite many, especially when there are examples like the Benveniste case.

Perhaps a better approach would be to try to understand the subject through research aimed at recognizing what it's all about? Dr. Dean Radin, a leading researcher at the Institute of Noetic Sciences (IONS), became interested in Emoto's research. In 2005, Radin and Emoto collaborated on an experiment in the hope of making Emoto's observations more reproducible for other researchers.

The experiment tested the hypothesis that intention directed toward water from a distance is reflected in the ice crystals (snowflakes) formed from the water. So, if water is thought of with love, it should appear as different snowflakes compared to when negative thoughts are directed toward it. Thus, for three days, a group of over 1,900 people in Austria and Germany directed their intention towards water samples that had been taken to an electromagnetically shielded room in California, USA. In the room, there were also control samples that the people sending intention did not know about, and additionally, there was a control sample of water placed outside the shielded room.

Ice crystals were made from the water samples according to the following process:

- Dean Radin purchased six bottles of Fiji water from a store in California, the same type used in a previous pilot study. He labeled the bottles A–F randomly by rolling the dice.
- Another researcher took the bottles to a laboratory, where two bottles were randomly selected to be treated, two were set aside as control water samples, and two bottles were placed outside the shielded room as samples. The information about which bottle was where was sealed in two envelopes, which were opened after the experiment. One envelope was left on his desk, and the other was taken to Radin's desk. He took the bottles to the shielded room, placing the treated water bottles on the table and the control bottles under the table. The control sample to be placed outside the room was placed in a styrofoam box on the top shelf of a bookshelf on the second floor of the laboratory building. Finally, he took pictures of the bottles' locations. The picture of the treated bottles without the control bottles' picture was sent to Emoto in Japan and his colleague Takashige Kizu. They later used the image to guide the intention of the remote healers to these water samples.
- Masaru Emoto conducted three separate sessions, one in Graz, Austria, and two in Germany, in Munich and Nuremberg, where a total of 1,950 people sent positive thoughts to the water samples. Emoto first showed the location of the IONS laboratory on a map at each session and then showed

a picture of the treated water samples. The picture also included the text "prayer for water." Then Emoto led the group in a prayer lasting about 5 minutes. The process was the same at each session.

• After the last session of the day, another researcher from California collected the bottles, wrapped them in aluminum foil, placed them in a box, then into another box with padded styrofoam, and the bottles were sent to Emoto's laboratory in Tokyo. At this point, Emoto and the other researcher in his laboratory learned about the control water samples that had been under the table next to the treated water samples. The Japanese researchers and Radin did not know which bottles belonged to which group; only Radin's colleague did. He did not participate in the research at this stage in any way.

• When the water samples arrived at Emoto's laboratory, his colleague took 50 drops of water from each bottle and placed them in 50 petri dishes. A cover was placed over each dish, and the letter (A–F) from the bottle was written on it.

• The dishes were placed on a tray in random order and then placed in a freezer, where they were kept at -13 to -22 °F (-25 to -30 °C) for at least 3 hours. The random arrangement on the tray ensured that any temperature variations in the freezer would affect all types of dishes equally randomly.

• Emoto's colleague moved the dishes from the freezer to a refrigerator room, which was constantly kept at 23 °F (-5 °C). In the room, he photographed the peak of each ice crystal using a stereo microscope at either a magnification of x100 or x200, depending on the presence and size of the crystal. Based on previous pilot studies, some water drops were not expected to form observable crystals.

• All 300 photographs, 50 from each of the six bottles, were then sent to Radin, along with a note indicating which bottle (A–F) and which sample number 1–50 each was.

• The aesthetic quality of the images was then evaluated by an impartial jury of 2,579 people. Each person evaluated 50 images randomly selected by a computer from a pool of 300 images based on *beauty and captivation* using a 7-point scale (e.g., not at all beautiful to very beautiful). Based on the previous pilot study, both beauty and captivation were asked about, as beauty is highly subjective. It can also be influenced by the method of photography, image quality, contrasts, and other factors. More images are often selected for the captivation category. However, the researchers did not want to change the category completely from the pilot study.

According to the results, the ice crystals that were treated, meaning those to which positive thoughts were directed, were more beautiful than the control group's crystals ($p = 0.03$, which indicates statistically significant results, the figure is smaller than 0.05, which is considered the threshold). Water samples treated with positive thoughts and words, such as love, peace, gratitude, wisdom, truth, angel, eternal, were also more symmetrical than water samples to which negative thoughts and words were directed. The result confirms the findings of a previous (double-blind) pilot study.

In this study, some of the shortcomings of previous experiments were eliminated, such as the fact that the control samples were now close to the treated samples, so that the difference between the samples could not be explained by environmental variations. The petri dishes were also placed randomly in the freezer to eliminate the effect of temperature variation. Researchers attempted to eliminate their expectations of the experiment's outcome by concealing various stages of the experiment. Image processing was also attempted to make the images uniform so that quality would not affect the results. By using an impartial jury, it was ensured that the selection of images was objective.

I described the stages of the experiment relatively accurately above because I wanted to highlight how challenging it is to study phenomena like these. Researchers attempt to study them in the same way as everyday phenomena have been studied for the past couple of hundred years. However, our understanding of observers' effect on the outcome of the experiment has changed significantly in recent years. How should the influence of a conscious observer be taken into account in these experiments?

Radin and Emoto discussed in their research publication how it would be possible to reduce the stages involving *"unintended intentional"* human choice and influence. Such stages in the experiment included, for example, selecting bottles for the experiment and different groups, selecting and preparing water drops in petri dishes, taking them to the freezer, and selecting ice crystals for imaging. When consciousness is an essential part of research, it cannot be fully determined how and in what way it affects the results. Of course, these stages could be automated. Bengston, in his publications, asks whether it is ever possible to form two separate groups, as inevitably the groups affect each other. In Bengston's experiments, mice also began to heal in the control group where someone from the treatment group visited only briefly. A unifying link was formed between the groups. How should water be interpreted in these experiments? Can it ever be considered completely separate from us, when such a large part of our body is water?

Emoto's rice experiment is even more familiar to many than the ice crystal study I described above. Many have also conducted the rice experiment themselves. I described my own experiment in my book *From Quantum Physics to Energy Healing*. The idea of the rice experiment is very similar to the ice crystal experiment. Negative words and emotions cause the rice placed in a jar to become moldy faster than the rice to which positivity is directed. However, the rice placed aside completely, to which no attention is given at all, molds the fastest. It continues the process it was set into. I recommend conducting this experiment – I was personally surprised at how impressive it is to see first hand what really happens in the experiment.

In the ice crystal experiment as well, the water sample that was completely disregarded formed beautiful crystals, albeit not significantly more beautiful than those formed from treated water. Is this also about the fact that when human interaction doesn't interfere with development, those processes continue as they were determined to do?

~

Austrian researcher Alois Gruber states in the movie *Water – The Great Mystery* that if you think negative thoughts, you simultaneously negatively charge yourself and also affect the water in your body, which constitutes 75–90% of you.

It also appears that each of us has our own unique influence on water. Professors Bernd Kröplin and Regine C. Henschel (Institute of Statics and Dynamics of Aerospace Structures, Stuttgart, Germany) conducted an experiment in 2017, in which four students were given pipettes and asked to fill them with water from the same glass. Each student then dropped a drop from the pipette onto a sample dish.

When examined under a microscope, it was found that although the water was from the same glass, and all drops were the same size, the drops were all different. Kröplin and Henschel concluded that each student's "energy" changed the drops.

Water seems to be able to preserve our intentions and also contain a trace of us.

"By holding the intention of peace towards water,
by thinking, speaking and acting with the intention of peace
towards water, water can and will bring peace,
to our bodies and to the world."

– Masaru Emoto

CHAPTER 7. HOW WE MAY AFFECT WATER

"Just as water turns solid
at the freezing point,
could matter be consciousness solidified
at the point of observation?"
– Johanna Blomqvist, *Hyperreality*, p. 111

~

Thoughts and emotions are known to have an impact on physical reality. In fact, we may even conclude that our physical, material world is also just consciousness. Water, in this sense, would also be considered consciousness. Consequently, the idea of influencing water would feel more natural.

CONSCIOUSNESS AND MATTER INTERACTING

One of the most significant and longest running research series related to the interaction between consciousness and matter is the PEAR (Princeton Engineering Anomalies Research) laboratory's consciousness research conducted from 1976 to 2008. PEAR laboratory was founded by physicist Robert Jahn and laboratory engineer Brenda Dunne. At PEAR, researchers studied the interaction between human and machine, conscious and unconscious influence, and remote viewing. Remote viewing involves seeking information by observing distant objects and events. PEAR is one of the most significant research projects of the late 20th century and has had a revolutionary impact on understanding the limits of consciousness. The significance of PEAR laboratory's research is probably not fully understood: it could even be considered the beginning of an entirely new field of science.

Millions of experiments were conducted at PEAR which individually might be seen as anomalies. However, collectively they point towards the same

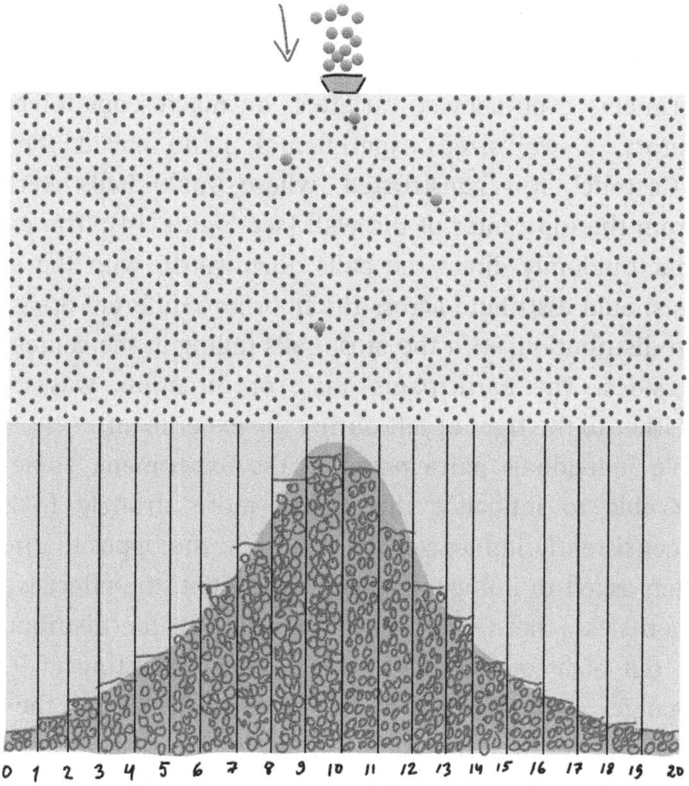

Figure 14. "Random Mechanical Cascade" experiment setup. The balls should statistically distribute into different compartments, collecting bins arranged across the bottom, according to the Gaussian distribution

narrative of the interaction between mind and matter: our minds influence the material world.

One interesting theme in PEAR's experiments was the significance of emotions on the results. In some experiments, the unconscious influence in a group was studied. Participants were not asked to influence the devices; rather, the experimenters only monitored the data produced by random event generators (REG). It was observed that the data deviated from randomness most in situations where there was a strong emotional connection among the participants, such as during group meditations in sacred places, meetings of close-knit groups, emotionally charged group rituals, and music and theater performances. There was a clear difference in the data compared to that corresponding to official seminars, conferences, and academic meetings.

The experiments even showed an effect on macroscopic objects. In the PEAR laboratory, a wall-sized experimental setup called the "Random Mechanical Cascade" was constructed, where 9,000 balls were dropped alternately from the top center of the wall (see Figure 14). The wall is filled with pegs that randomly direct the balls downwards, with the expectation that the balls should distribute across the floor into different areas according to a Gaussian distribution, with the most balls falling in the center and fewer towards the sides. The participants' task, or intention, was to shift the distribution either to the right or left during the experiment.

Twenty-five individuals participated in the experiment, some of whom were clearly able to influence the results more strongly (4/25). Some participants consistently influenced the results in the opposite direction. For example, when asked to influence towards the right, the effect was towards the left. Nonetheless, there was a clear deviation in the distribution in the experiments: out of a total of 3,393 drops, each consisting of 9,000 balls, significant changes were observed in 15% of them. In 63% of the drops, the deviation was in the desired direction. An interesting observation was that it was more likely to shift the distribution of balls to the left rather than to the right in the experiments. Researchers couldn't explain why this asymmetry occurred, as there was nothing in the experimental setup that could explain this.

Wishing for good weather

Roger Nelson tells of an interesting observation at Princeton. Weather greatly influences what we do and plan. Princeton University regularly hosts a fall opening event attended by thousands of alumni, graduates, their families, and university staff. The event is always held outdoors, and it's often noted how excellent the weather is during the event. When acquaintances meet, they remark on how nice the weather is again, as always. Comparing the statistical rainfall in Princeton and the surrounding areas, it has also been noticed that on those particular days, Princeton experiences significantly less rainfall than usual for Princeton or its vicinity.

Usually, such observations are difficult to study or compare, but data has been collected at Princeton for the university's three largest events for decades, some of it dating back 250 years. According to the statistics, those days have indeed been significantly less rainy than the surroundings. Nelson suggests that in the future, we should collectively wish for good weather, as besides extreme luck, the results can only be explained by the collective influence of the participants.

Dissolving clouds

In addition to wishing for good weather, one can also experiment with influencing water by dissolving clouds. I first read about this in Bill Bengston's book *The Energy Cure*. Of course, I had to try cloud dissolving myself.

Try cloud dissolving - these are my instructions on how do it:

- Choose a partially cloudy day and select one of the clouds drifting in the sky. I started with a fairly small cumulus cloud on a sunny, partially cloudy day when there were quite a few clouds in the sky.
- Mentally establish a connection with the cloud. Feel how you are one with the cloud. Look at it, sometimes closing your eyes and still feeling the connection to the observed cloud.
- Visualize in your mind's eye how the cloud begins to split in half or gradually disappears entirely, like smoke into the air. Make this visualization in your mind as real as the image of the cloud you see in the sky.
- Let go of the need to change anything. You have already created a mental image of how the cloud changes and disappears; now give it the opportunity to happen.

I practiced in my mind for a moment, and at some point, I dared to open my eyes, and to my amazement, the cloud had started to shrink! Pieces had already detached from its edges, and it was clearly smaller. Over the next few minutes, the cloud shrank even further until it eventually disappeared.

What exactly happened? As I looked at the other clouds, they still looked almost the same as when I started. Why did that particular cloud I focused on disperse? Naturally, my mind tried to interpret the event as a coincidence, reasoning that perhaps there were more air currents at that point that dispersed the cloud, and so on. Of course, there was also a feeling that I should try again.

By the way, it's likely that the second time won't go as well due to beginner's luck. The saying comes from the fact that beginners often perform better than others, for example, sinking a golf ball into a hole or getting good cards in a card game. This phenomenon has also been observed in studies, such as those conducted by the PEAR laboratory, where in experiments investigating the interaction between humans and machines, it was found that participants were generally able to influence the machines better in the first series of tests compared to subsequent ones. Other test series were almost always weaker, and in the third series, the results were similar to the first series but not as good.

Rain dances work in the same way as cloud dispersal. Instead of dispersing clouds, the goal is to create more clouds and, of course, rain, for example, to end a dry season. The tradition dates back to ancient Egypt and is particularly remembered as a ritual of Native American tribes. The rain dance ritual has been passed down orally as tradition. Often, rainmaking is called on to restore natural harmony. When a rainmaker observes the prevailing weather conditions and how they affect the community and themselves, they turn inward to restore their own balance first, thus restoring harmony and, along with it, rain.

In ancient Rome, there was a ceremony called *aquaelicium*, which meant calling for water. The ceremony used a special stone called *lapis manalis*, which enhanced the flow of water, with "*manalis*" coming from the Latin verb "*manare*" – to flow. Sacrifices were made to Jupiter, the god of the sky and its phenomena, and water was poured over the stone. Stones seem to be part of many water ceremonies. For example, in my home country Finland, on Jaakko's Day, or St. James' Day, July 25th, according to an old saying, Jaakko throws a cold stone into the water and the waters start to cool. Jaakko, or Saint James the Greater, is one of the Twelve Apostles of Jesus, and the stone is also often said to be a symbol of sins and it's thrown away as a sign of liberation from sins. Also traditional in Finland is the sauna, where water is poured over heated stones to create steam and intensify the heat.

~

Dean Radin, the leading research scientist at IONS, says that they at IONS are interested in the interaction between the mind and water because we are mostly water: *"Perhaps water could also provide an answer to how energy healing, distant healing, or other healing methods related to the mind and consciousness work – perhaps the mind can affect the properties of water, which would naturally also be reflected in us."*

As I described in the previous chapter, Masaru Emoto's ice crystals fundamentally deal with the same phenomenon – how water has the ability to react to thoughts and emotions. Positive thoughts affect water, and conversely, such water could positively affect us as well.

Since I became acquainted with Reiki and energy healing, I have been *"channeling energy"* to various targets. In addition to treating humans, animals, plants, and trees, I have also treated inanimate objects such as washing machines, car windshields, and as a busy mother, even grocery bags – then I don't have to remember to treat or bless the finished food later, as all the food has already been treated while in the grocery bag.

Although I speak of *"channeling energy,"* I actually mean a change in thinking. Energy is not actually sent anywhere. Instead, there is a change in the idea of what is possible and real. Channeling energy actually means bringing the desired state of reality into this moment through intention. What is best for the recipient is treated. The future already exists because, in reality, there is no time or distance; they are just experiences. In the Reiki method, a change in thinking is already included in the principles of Reiki: *"Just for today, be happy, expect the best, be honest, be kind, be grateful."*

As I mentioned in the chapter of homeopathy, treated water seems to retain an intention to heal. Bill Bengston has also observed a similar effect with cotton and wool. It appears that there are several ways to "preserve" a healing intention in both organic and inorganic materials as well as in water. I will now tell you about some studies where the connection between healing intention and water has been examined. Perhaps these studies could reveal something new to us about water and ourselves? Since about 70% of the human body is water, studies on the effects of energy healing and other methods on water actually also tell us about the effect of water on us. If changes in water's previous state are visible, can differently treated "charged" water similarly affect us?

HEALED WATER AND PLANTS

Many Reiki and energy healers personally notice the effect of treatment on plants. One particular Reiki client stands out in my memory, whom I visited weekly to perform Reiki treatments. After one session, she asked me if I had noticed what had happened to the plant in the room where I had done the treatment. At the beginning of the treatment, she had intended to throw the

Figure 15 a and b. Photos of nasturtiums taken monthly. Treated flower in front at the bottom.

plant away as it was in poor condition despite all the watering and fertilizing efforts. However, the plant had now started to grow – almost in the middle of the darkest time of the year. It had grown new light green leaves and had suddenly started blooming! She hadn't changed anything in her plant care routine. The only change was the energy healing that we were now doing weekly near the plant.

One autumn day, I decided to treat the nasturtium flower (also called Indian or monk's cress) in the flower box in my garden. There were many flowers in the box, but on a whim, out of curiosity to study the effect of the treatment, I chose one flower: I treated it every time I passed by that day. The days grew shorter, and the nights grew colder. There were even frosty nights, and all the other flowers around that treated flower wilted one by one, sagging and dying from the cold, but that one flower just bloomed. It looked unusually bright orange standing there amidst the other wilted flowers, facing towards the sun. It finally wilted one night after I had shared the story on Instagram with my social media followers – as if acknowledging that its mission was now fulfilled and it was time for it to go.

I talk about the treatment of nasturtium seeds in my book *From Quantum Physics to Energy Healing* as well. I planted 32 seeds in two jars, treating one and setting aside the other to compare the effect of the treatment. The effect was small but clearly noticeable. First, all the seeds (15/15) in the jar I treated germinated within 14 hours of planting, while in the control group, 12 out of 17 seeds germinated. The seeds in the treated jar also grew faster: by the time 72 hours had passed since planting, all the plants in the treated jar had stems over 1.18 inches (3 cm) tall (15/15), whereas in the control group, only 9 out of 17 seeds had stems over 1.18 inches tall. On average, the plants watered with treated water grew faster and taller. This was, of course, a single series of measurements, but the results were consistent with other studies investigating the effects of energy healing on plants.

Journalist Lynne McTaggart, along with Professor Gary Schwartz, conducted experiments with plant seeds and water during one of her lectures. She showed the audience a picture of a water jar and asked them to send the water an intention that "all seeds given this water in the jar would grow to at least three inches (2.57 inches = about 7.5 cm) tall within four days of germination." When the seeds were then watered with the treated water, they indeed grew better than seeds watered with regular water. After measuring the sprouts four days later, the "intention-treated" seeds were on average 1.88 inch (4.77 cm) long (with 30 seeds) compared to seeds watered with regular water (with 90 seeds), which were on average 1.83 inch (4.66 cm) long. The difference was 0.04 inch (0.11 cm), which however is not yet a statistically significant deviation. Similarly, in my own experiment, the intention-treated plants grew taller.

In McTaggart and Schwartz's experiment, an interesting observation was made: typically, not all seeds germinate, and this was the case here too, with up to 10% failing to germinate. However, all seeds treated with intention did germinate. This was also the case in my experiment. Naturally, these experiments should be repeated to draw any final conclusions.

As for whether there is a difference between treating the plant or the water used to water the plant in the experiment, I believe there shouldn't be. Plants are very water-based. In nature, there are many plants that are particularly water-based, such as watermelon (92%), cucumber (97%), and aloe vera (99%). Aloe vera, in particular, contains a lot of EZ water, according to Gerald Pollack's research. This may explain why it is healing to burned or injured skin.

The potential effect of healing intention on water has actually been experimentally studied since the 1960s. Dr. Bernard Grad of McGill University in Montreal conducted many well-documented studies at that time, in collaboration with Hungarian healer Oscar Estebany, relating to plants, water, and healing intention. He studied the effects of energy healing on features like increased plant height, chlorophyll content, leaf quality, and growth rate. Based on the studies, he concluded that seeds and plants watered with treated water grew taller, stronger, and overall healthier than those in the control group. The experiments were repeated several times with similar results.

Similar conclusions have been drawn in later studies, such as Dr. Robert Miller's on the effects of energy healing on plant growth rate. The Spindrift group also found a similar effect on plant growth rate in studies investigating the effects of prayer on plants. According to their research, plants prayed for always grew faster and healthier, even under the same growth conditions.

The effect was independent of the distance between the healers and the plants, extending even kilometers away.

Shiah and his team studied the effect of water treated with intention on the growth of Arabidopsis thaliana, or thale cress. This fast-growing plant in the mustard family is found especially in Africa and Eurasia and was the first plant whose genome was sequenced. Three Buddhist monks treated the water used to water the plant seeds. The control was the same water, but untreated. The experiment focused especially on the growth of leaves and the content of anthocyanin and chlorophyll. Anthocyanins are plant flavonoids with red, blue, and purple pigments found in the plant cell sap. Chlorophylls are the green pigment the plant needs for photosynthesis. Leaf growth and anthocyanin content showed significant changes: treated water was clearly better, although there was only a slight improvement in chlorophyll content.

Shiah's group also studied the effect of treated water on human mesenchymal stem cells – cells isolated from the bone marrow that can differentiate into bone, cartilage, and fat cells. In a publication released in EXPLORE in November–December 2022, they state that based on double-blind studies, treated water seems to affect the growth, properties, and aging of mesenchymal stem cells. The conclusions in the publication warn against making far-reaching conclusions based solely on this one study; even though it is a double-blind study, more research is needed. However, it must be said: what works for plants likely works for animals and humans too.

While I haven't provided a comprehensive summary of the effects of treated water on plants here, all the results consistently show that, although the differences in experiments seem small, treated water appears to accelerate growth and make plants healthier. I, thus, decide to test the effect of treated water on the growth rate of plants myself (see Chapter 9), too. Roney-Dougal's research group, for example, concluded, in a publication studying the growth of lettuce, that plants watered with treated water eventually grew 10% larger than those watered with regular water. William Bengston, whose experiments with mice I mentioned in my book *From Quantum Physics to Energy Healing* investigated the effect of treated water on the growth of tulips in experiments conducted in the Netherlands. Treated water was added to the plants via the irrigation system, and the effect was clearly visible: flowers watered with treated water grew faster and larger than untreated ones.

Individual experiments may be anomalies, but together they indicate that this is a statistically significant phenomenon that should be taken into account or further investigated. I will definitely continue to treat my plants with energy healing when planting them and also treat the water used for watering plants. Nothing is preventing us from testing for ourselves the effect of treating water and plants. I suggest that you too test what impact it has. There's no harm in trying it out.

If energized water affects plants, it must also affect us. Perhaps it will be shown that the quality of food depends on the mood and well-being of the person who grew and prepared it, and the value of well-being will be increasingly appreciated?

WATER STORES HEALING INTENTION

In my book *From Quantum Physics to Energy Healing*, I discussed the groundbreaking energy healing research conducted by Dr. William Bengston. It was the reason why over a decade ago I became interested in energy healing and Reiki and began to learn about them.

In Bengston's studies, mice were injected with a highly aggressive form of breast cancer, and for the first time, the mice recovered from it. In over 2,000 previous experiments with this form of breast cancer (mammary adenoma cancer), all mice had died before the 27th day from exposure. In Bengston's experiments, over 90% of the mice actually recovered. The results of this study have been replicated in 16 experiments conducted at 10 different universities to date.

Currently, Bengston is researching the possibility of "preserving" the healing effect (I'll use the term 'treatment' or 'healing' below) and also how to expand the effect to a larger group of recipients, as there really is a need for treatment. Scaling and the ability to preserve the effect are needed for energy healing to become a common and widely used form of treatment someday. Bengston's view is that anecdotes of healing cases can be convincing but alone are not enough for energy healing to become as widely used a treatment modality as those considered within the sphere of Western medicine.

According to Bengston's research, treatment can be stored in both organic and inorganic materials. When blood from mice that had recovered from cancer was transferred to mice still suffering from cancer, those mice also recovered. A significant effect on cancer cells was also observed in experiments where only cell fluid was treated.

In his experiments, Bengston found that water can also retain treatment, as can cotton. In Bengston's mouse studies, even cancerous mice that received water treated with energy healing recovered from cancer – without being treated themselves using the method. These more comprehensive studies were needed since there was a lot of anecdotal evidence, or cases where people reported healing from leukemia or irritable bowel syndrome (IBS), for example, after drinking treated water or reported the disappearance of localized conditions such as cancerous tumors using treated cotton.

By studying treated water and cotton, answers can be obtained to the question of whether treatment can be retained, but it is also necessary to investigate the scalability of the treatment. To investigate this, Bengston, together with a Dutch group, designed an experiment similar in process to homeopathy: water was first treated with Bengston's developed energy healing method, after which the treated water was diluted, shaken, and then diluted and shaken again.

They settled on dilution because they had heard from some researchers that the dilution process can increase the potency of the liquid from the original. Curious and skeptical, they examined the process and got encouraging results: it really seemed that such potentization was related to the effect. Bengston and his group tested different dilution methods and eventually settled on their chosen "recipe" after consulting with an experienced clairvoyant friend. Through his intuitive advice, they found the right dilution solution. Although dilution is also discussed in homeopathy, Bengston and Nies emphasize that their solution is not homeopathic, even though water underwent dilution, and the dilution process included vigorous shaking of the liquid.

The treated water was diluted with untreated water – that is, water was diluted with water, and between dilutions, water diluted with water was shaken. As Bengston notes, the process seems crazy, but nevertheless, it seems to make the water stronger. He has no theory as to why this happens. It seems that at the moment, no one else has an answer as to why dilution and shaking achieve this effect.

The diluted liquid was first tested in preliminary tests, and the results were very positive. For example, Parkinson's patients who received diluted treated water reported more significant improvements in their condition than when treated solely with energy healing methods. Patients with neurological conditions or bipolar disorder reported dramatic and even permanent recoveries. Inspired by such anecdotes, Bengston and Nies decided to conduct a more comprehensive study in which diluted treated water solution was given to volunteers for eight weeks. Three clinical pilot studies involved a group of volunteers, a total of 86 people, with conditions including cancer, arthritis, depression, anxiety, pain, and Parkinson's disease. Participants took two drops of the water solution under their tongue four times a day. At the start of the trial, 90% of participants reported their condition as severe, with 22% having cancer, for example.

Participants were asked to answer questions about their well-being every two weeks, and at the end of the study, to assess their overall progress and whether any changes had occurred.

The result was that 98% reported better health, and even significant physical, mental, and spiritual well-being. The results were not only minor; they were even significant. Some effects seemed even stronger than in local energy healing sessions, where the patient receives energy healing from the practitioner in the same space, as in the case of Parkinson's disease. Overall, the physical, mental, and spiritual well-being index was at its highest level after eight weeks. Many participants would have wanted to continue taking the water drops even after the trial! Bengston has packaged the water drops together with a Dutch group, and Bengs drops can now also be purchased.

I have sometimes been asked, usually during the Olympics, World Championships, or other major sporting events, whether energy healing should be given to elite athletes, for example, to improve their results. I have answered, absolutely! This answer is also supported by the pilot study conducted by Bengston and Nies, which also tested the effect of treated water on athletes' performance. Weightlifters reported breaking their bench press and deadlift records, cyclists pedaled faster and farther. In short, athletes gained more strength and speed, recovered faster from their performances, and even their oxygen intake improved!

Bengston's conclusion from the research is that something significant indeed happens: The process he describes, treating water by diluting and shaking it, brings about changes. Preliminary studies have also yielded significant results for COVID-19 patients. The next question is: What if it turns out that we have a functional, scalable treatment method for many ailments, one that doesn't even have side effects because it's just water? The treatment would be something truly unconventional, not at all what our assumptions about treatment likely are. Even though double-blind studies and others have been conducted, based on previous cases, one can well anticipate the reception these studies might receive: requests for repetition, continued scrutiny for errors. For an ordinary person, if indeed it's a treatment method without side effects or risk of overconsumption, these results are nonetheless revolutionary. I'm sure many would be willing to try such a treatment – after all, there's no possibility of harm since it's just water, and it could potentially change someone's life!

Bengston is further researching the subject, aiming to determine the extent to which the effect is due to the placebo effect and what is due to *resonance bonding*. Placebo refers to the tendency to improve even with any treatment as long as the patient believes the treatment is effective. Resonance bonding is a term Bengston uses to describe how it can be difficult to distinguish between two groups when some form of connection has been established between them. In Bengston's mouse experiments, even the mere presence of the same practitioner visiting both groups seemed to be sufficient for the mice in the control group to start recovering. In a way, this is quantum entanglement at the level of everyday life. Once two groups, particles, or anything else have been together, the information between them still exists.

Bengston has also investigated, together with Sarah Besemen, Dean Radin, Michael Turner, and John McMichael, the storage and preservation of healing intention, somewhat similar to Montagnier's research. They recorded electromagnetic signals while three energy healers treated cotton. When the recording was played to cancer cells, significant changes occurred in the cells' genes, indicating changes in gene expression.

AFFECTING WATER CONSCIOUSLY AND UNCONSCIOUSLY

Researchers Helané Wahbeh and Dean Radin, along with their team at IONS, conducted a pilot study examining the effect of energy healing on hand and wrist pain. The study involved 193 participants and 17 energy healers. Among other things, the study followed the participants' expectations of treatment, pain experience, heart rate variability, synchronization of the heartbeats of the healer and the recipient, and gene activity. Changes in the environment, such as water pH, conductivity, evaporation, and quantum noise – changes in environmental entropy – were also monitored. Previous studies have noted that when intention is focused on a single target, changes can be observed, including in data generated by random number generators, as if reality becomes more organized in line with the intention. The studies found significant reduction in pain experience immediately after the treatment session, and the effect lasted up to three weeks after treatment. Additionally, changes were observed in the data from random number generators compared to control conditions. Changes were also observed in water near the healer.

It appears that changes occur in the hydrogen-oxygen (HO) bonds of treated water or water kept nearby during treatment. In the study by Wahbeh et al., water was present near both the healer and the recipient. They noticed changes in the bond lengths and H-O-H angles of distilled water molecules. These changed during the experiment compared to control water, with the HO bonds stretching and the angle widening (Figure 16). In the "direct" test, healers treated water samples directly, while in the "indirect" test, healers and recipients wore aliquots attached to lanyards during treatment. Laboratory-grade distilled water and Fiji-brand bottled water were used in the experiment. The samples were analyzed using an infrared spectrometer. The average spectrum was used as a control, taken before and after the intention-healing period.

Interesting to note is that changes were seen only in distilled water, not Fiji water. This may indicate that minerals in the water have a significant effect. Changes occurred in the HO bond when water was treated directly or

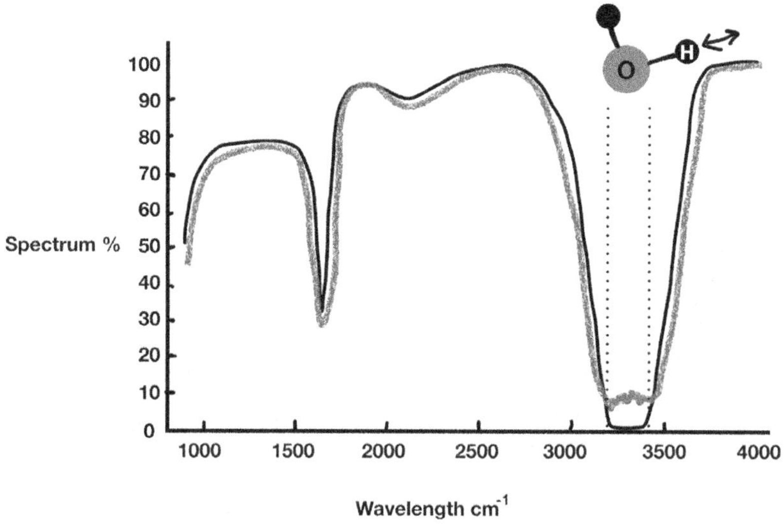

Figure 16. Spectrum of distilled water before (in black) and after treatment (in gray). Water samples were hanging around the necks of the healers during treatment. The wavelength range 3200–3400 cm^{-1} is associated with stretching of the HO bond (Radin et al.).

indirectly during the treatment while healers wore water bottles around their necks. No changes were observed in water bottles near the recipients.

Radin's group cannot explain why changes occur specifically near the healers and not at all near the recipients. However, a similar phenomenon was observed in another study: Bengston and Moga investigated the magnetic field around energy healers and recipients and found that magnetic changes occurred only near the healers, not the recipients.

Why did distilled water change then, but not Fiji spring water? This may be because distilled water is very uniform, whereas Fiji water as a commercial product likely contains more minerals and other substances that can affect the spectrum. This conclusion is supported by the IR absorption spectra taken before the intention experiment of distilled and Fiji water. The most significant differences in the spectra of the waters were in the HO bond wavelength range, and perhaps distilled water was more sensitive to change.

Minerals appear to have a significant effect on water, as noted in experiments with capillary tubes by Derjagin and in various dilution experiments. Water seems to change in interaction.

INTENTION EXPERIMENTS ON WATER

Journalist Lynne McTaggart regularly organizes intention experiments in which a group focuses its attention and healing thoughts on a common target. The instructions are for each participant to quiet their mind and individually focus on a specifically formulated intention statement for each target. This could involve focusing on the health of individual people, affirming that the person is healthy and thriving in every way. The intention is sent from the heart while listening to collective energy. McTaggart has organized large intention events where focus has been on peace in a specific area, such as in Sri Lanka, the most violent areas of Afghanistan, or in Washington, D.C. In the Washington, D.C. experiment, participants were asked to focus on reducing violence by at least 10% and on bringing peace to the area. In the months following the experiment, indeed, there was a decrease in violent crime – in some areas of Washington, D.C. by over 30% for months!

McTaggart has also organized intention events focusing on treating water. In one experiment, attendees at her lecture in Florida were tasked with sending love in their minds and causing water samples to "shine" in physicist Dr. Konstantin Korotkov's laboratory at St. Petersburg University. In pilot studies, this indeed occurred. Korotkov has developed the GDV (Gas Discharge Visualization) device, which measures, among other things, light from the water surface, emission activity, and pH. Surface emission activity of liquids refers to the surface atomic composition, such as the presence of hydrogen atoms that could form bonds. According to Korotkov, this could also explain how water can store and retain information. The GDV device has been found to be particularly sensitive to physical and chemical changes in liquids and capable of detecting changes that may not be visible in conventional chemical analyses. Experiments conducted with other lecturers by Korotkov have also shown changes in GDV device measurement data at the beginning and end of the experiment. The size of the group does not seem to affect the results: according to Korotkov's experiments, an experienced healer can produce an even more significant change in the data than a large group.

In a later experiment, McTaggart and Korotkov, together with Dr. Gary Schwartz's team at the University of Arizona, investigated water purification through intention. Schwartz's team took GDV images of tap water samples in four petri dishes and four mineral water samples. The difference in the images was clearly visible: The tap water dishes were cloudy, like pictures of the sun taken from behind clouds, and the mineral water images were clear. One of the tap water dishes was selected as the dish to send the intention for clean water, and the image was sent to participants in McTaggart's intention experiment. Their task was to make the tap water image more like the mineral water images. This indeed happened. After the intention experiment, new images were taken of the dishes. Upon examination, it was noticed that one of the tap water images was significantly brighter – exactly the one to which intention was sent!

The experiment was repeated later during a conference, but this time, a similar difference was not observed. There were no other differences from the previous experiment, except that this time the water was in the dishes for a shorter period, only a couple of days, rather than five as in the first experiment. Schwartz and his research team couldn't find a reason for the difference, and of course, the experiment should be repeated. Perhaps the results of the first experiment were also influenced by "beginner's luck" or chance?

Materials science professor Rustum Roy from Pennsylvania State University studied, with McTaggart, the effect of intention on water structure using Raman spectroscopy. Raman scattering, or the Raman effect, is based on the scattering of electromagnetic radiation, where the frequency of the radiation changes due to the influence of the substance it encounters, in this case, water. This phenomenon can be used to identify molecules and determine their structures, among other things. Once again, intention was directed towards the water in the laboratory, without the measurers knowing the timing of the intention. The measurement group observed significant changes in the water approximately an hour before the intention was supposed to be directed towards the water. However, the outcome of the experiment was found to be unclear. Similar changes before the actual start of the experiment have been observed in other experiments, including one of Korotkov's intention experiments. Changes often began even before the actual event in experiments conducted by the PEAR laboratory, where the influence of consciousness on the data produced by random number generators was studied. Perhaps the influence of consciousness cannot be examined with pinpoint accuracy in part because it is hard to determine when an intention begins in time?

Masaru Emoto proposed to Lynne McTaggart an intention experiment at Lake Biwa in Japan. The lake is one of the oldest in the world, and its water is used as drinking water for nearly 14 million Japanese. The water quality of the lake has significantly deteriorated over the last few decades. A water sample taken from Lake Biwa and a control sample were photographed using Korotkov's GDV device, and pH was also measured. The experiment was to investigate the effect of intention on the molecular structure of water in a similar way to the experiment conducted with Rustum Roy. McTaggart lectured on the topic in Japan and asked a live audience of 500 people and online participants to direct their intentions towards the water in Lake Biwa so that it would be once again like the water of a mountain stream, equally clear, fresh, and pure.

After the experiment, the water samples were measured, and there was a clear change in the parameters and signal representing the water. At the end of the experiment, the intensity of the signal remained strong, but its target area changed, indicating, according to Korotkov, a permanent change. The experiment was conducted in 2010. In measurements made in 2019 to determine the oxygen content of the lake bottom, the readings had improved compared to 2010 when the readings were at their lowest.

You can read more about intention experiments in Lynne McTaggart's books, such as *The Power of Eight*, where she introduces the method, and *The Intention Experiment*, and provides more detailed information about the experiments and observations. Up to this point (early 2024), 39 major experiments have been conducted, and of these, 35 have shown a positive, measurable, significant change. Collective intention, thoughts, and feelings have an effect. Positive intention is reflected in various ways, such as in the growth of plants, accelerating the germination and growth of seeds, purifying water, reducing violence in war zones and cities, and healing people.

McTaggart considers the feeling of unity experienced by those participating in intention experiments to be the most significant phenomenon. The sense of presence and connection is a particularly strong experience at these group events. The feeling can even be felt remotely, from a distance. I fully agree with this, as I have experienced the same in my remote healing sessions. I have regularly conducted remote healings and group remote healings for years, sometimes involving hundreds or even thousands of people. I will tell you more about remote healing later in this book.

McTaggart also makes the point that sending intentions reflects back to those who direct their intentions towards a common target. Participants have experienced their own lives becoming more loving and peaceful, and their own dreams starting to come true. Intention seems to be two-way. I have come to the same conclusion with energy healings: It's not just a one-way flow from the healer to the person receiving the healing; the connection works the other way around as well. The healer also receives healing while performing the treatment.

SUMMARY OF WATER STUDIES

In summary, it could be said that changes indeed occur in water in an anomalous way, influenced by intention, thoughts, and emotions! We can affect water, and of course, it also affects us.

In Radin's research group studies, it is mentioned that some kind of *"energetic influence"* is needed for changes in water, but they do not speculate on what this influence could be. The quotation marks around energetic influence are because it is not known whether it is a causal – cause-and-effect – phenomenon and whether energy is really energy, similar to what is discussed in physics. Phenomena related to the mind and consciousness cannot be examined in the same way as everyday events. It seems more likely that it is a phenomenon related to the transmission of information.

Radin's research group also speculates whether the observed effect could be an indirect influence, for example, the influence of some other environmental factor. However, this alternative is not supported by the fact that similar changes in water have been observed in other studies, such as:

- **Emoto's ice crystal experiments**, mentioned earlier in this book. I also recommend familiarizing yourself with water researcher Veda Austin's crystallographic photos. She photographs water in its 'state of creation', the space between liquid and ice. These photos seem to amazingly capture the essence of their object.
- **Microscopic studies examining water droplets treated with intention.** In their 15-year-long microscopic water study, Dr. Bernd Kröplin and Regine C. Henschel's group observed a clear difference between water samples from different environments. The proximity of cell phones, ultrasound, music, plants, rocks, and the influence of individuals' intentions were clearly visible in the microscopic images.
- **Studies on plants watered with intention treated water**, as discussed earlier in this chapter.
- **Survey studies collecting subjective mood assessments from individuals who drank treated tea or wine**. In Shiah & Radin's study, oolong tea treated with positive intention produced a better mood than ordinary, otherwise identical tea. If the individual also believed they were drinking treated tea, their mood improved even more, but only if they actually drank treated tea.
- **Laser light and light scattering measurements investigating the effect of intention on water structure.** Pyatnitsky & Fonkin concluded in their studies that human consciousness seems to affect water structure. Changes in the water structure were more evident than when measuring properties of water dependent on structure, such as conductivity, surface tension, elasticity, density, or light refraction.
- **IR absorption measurements examining changes in molecular bonds.** In Schwartz et al.'s 2015 study, they replicated the methods of previous studies, paying particular attention to blind testing, randomization, and attempting to eliminate possible sources of error. The study measured IR spectra of water samples near 14 healers and 14 recipients. Each healer used their own preferred energy healing or intention method. Changes in IR spectra were observed particularly near the healer. Changes occurred in the wavelength range associated with the water OH bond and its stretching. The study also noted that the effect of treatment

did not appear to depend on time: if the interaction lasted longer, the effect did not strengthen.

As noted in previous chapters, the effect is not dependent on distance either. Several studies support this conclusion:

- **LED/laser measurements, where water reacted before the experiment.** In Kernbach et al.'s study investigating long-range effects between devices and human intention, it was observed that a receiving sensor reacted indirectly to LED light radiation even through illuminated water for half an hour to 72 hours before the experiment. There were about 900 measurements in total. They also observed how the emotional state of participants in the experiment had a significant impact. The stronger and more intense the emotional state, the stronger the sensor reactions. A relaxed and calm emotional state did not affect the sensors, nor did mental focus on the sensors.
- **Raman spectroscopy measurements investigating the effect on water from a distance.** Qi Gong master Yan Xin sent *Qi* to a water sample seven kilometers away, resulting in a clear change in bond lengths and angles.
- **GDV (Gas Discharge Visualization) device measurements (Korotkov's equipment).** K. Korotkov has been studying remote influence on water samples for over 20 years. Statistically significant changes have been observed in numerous experiments; the healer appears to make the water "glow" in the images taken by the device. In one experiment, German healer Christos Drossinakis treated water samples in St. Petersburg for 10 minutes from Japan. A significant change was observed in the sample images on the days when he sent treatment, but no change was observed in the control samples.

Water is a very sensitive fluid. Its structure is constantly changing because everything affects it. Water changes in interaction. We influence it unwittingly. We affect it, as does our entire planet, the sun, and the entire cosmos. For good and also for ill, everything has an impact.

CHAPTER 8. THE MYSTERY OF HOLY WATER

"Water is the mirror of nature."

– Francis of Assisi

~

I feel I've developed a very special connection to water. Water seems to "follow" me everywhere and "announce" itself in different places. I notice water and things about it in various places. Water is also present in my dreams. It seems to play an increasingly larger role in the lives of people close to me, as if it "attaches" to them.

One October evening, I was teaching at a Reiki 3B course, where the participants learn Reiki, the method of spiritual growth and natural well-being. The Reiki 3B course is a teacher training course in Reiki, at the end of which participants become so-called Reiki Masters or Reiki teachers. To start the evening, we did a meditation, and I also performed a revision activation for them. Activations, also known as initiations or attunements, are a characteristic part of Reiki, through which a Reiki teacher connects Reiki practitioners to the network of all Reiki practitioners. The purpose of activations is to help and accelerate the ability to channel energy. In the recap activation, the teacher strengthens the student's connection to Reiki and its source. Reiki activations are always received from one's own teacher, with the Reiki lineage, network, leading through teachers, ultimately to the gatherer and founder of Reiki knowledge, Mikao Usui in the 1920s.

The activation was a very special experience for me as a teacher as well. My students were seated in a circle. I walked in front of them, behind them, sometimes placing my hands on their shoulders, and performed the activation as I had learned from my own teacher. It felt as if the entire material world disappeared as soon as the activation began, and we transitioned into light beings in an infinite dimension, yet still in a familiar and safe space filled with love, like home. Occasionally, some moments came to my awareness:

how I moved from one light being to another and how I felt a strong vibration, a sensation of vibration with each one. This vibration was strengthened during the activation. I felt I was helping each person open up, as if straightening their spine. Many of those being activated shed tears; some kind of purification was taking place. The water element was very strongly present. When we shared our experiences of the activation, someone mentioned seeing a water-like blue, another felt waves, and someone else described "diving from deep water to the surface to take a breath." Hearing this, I realized that in all the Reiki sessions I had done in recent weeks, water had been present in some way through experiences like these. In one remote healing session, for example, I experienced I was by a calm lake on a pier. I saw the landscape very clearly and described it to my client. During the session, they mentioned arriving at their cottage and looking at the lake by the shore and on the pier. They were looking at exactly the same landscape I had described.

BLESSING OF WATER

After completing the Reiki teacher training course, the same atmosphere continued the next day in the basic Reiki course: Water followed along. I had reserved a space in a cultural center for the course. They had several different-sized classrooms, an auditorium, and corridors filled with books. Often I had held courses in the same place, but this time my feet led me in a different direction than usual from the stairs for some reason. In front of me opened an unknown, dark corridor lined with long bookshelves. I turned on the lights in the corridor and looked ahead. In front of me was a glass display case, and the title of the poster attached to its wall caught my eye immediately: Blessing of Water. Cold shivers ran through me.

I read from the text how the blessing of water is a ritual in the Orthodox Church. It can be performed near natural water or at a well, in addition to home or church. There are both a grand and a minor blessing of water, in both of which the water is sanctified by the power of the Holy Spirit. The grand blessing of water is performed only once a year, on Theophany, the sixth of January, the feast day of the Baptism of Christ, in church or by natural open water, while the minor blessing of water can be done several times a year on different occasions.

In the display case was a picture showing a metropolitan bishop holding a cross in his hands and another priest sprinkling water around him with an *Aspergillum*, "holy water sprinkler," while holding a vessel. In addition to these, the blessing of water requires a ritual censer and a Gospel book. The actual sanctification of the water takes place by immersing the cross in it.

In the book *Of Water and the Spirit*, Orthodox theologian Alexander Schmemann explains that the blessing of water begins with an initial blessing, followed by specific petitions asking the Lord to sanctify the water with the power of the Holy Spirit: It asks that the water be made a protection "against the averting of every snare of enemies both visible and invisible."

Schmemann writes: *"It is through his enslavement to the world and its matter that man became the slave of the demonic powers. The liberation of man begins with the liberation, i.e. the purification and the redemption of matter, its restoration to its original function: to be a means of God's presence and, therefore, to be a protection and defense against the destructive 'demonic' reality" (Of Water and the Spirit, p. 42).*

So water is a means by which man is freed from matter. The circle is closing. I begin to understand why all of this came to me now as I write about the interaction between consciousness and matter and how we can affect water.

The blessing of water continues with the priest praying first for himself, that he may interpret and act correctly. I recall Søren Kierkegaard once said that the purpose of prayer is not to influence God, but rather to change the nature of the one praying. After prayer, the priest turns to the water, where he is *"as if facing the whole cosmos on the day of creation"* (*Of Water and the Spirit, p. 45*). This is followed by a prayer of blessing and sanctification of the water, thanking God for the whole world, as gratitude is the most important act of man, through which he fulfills his task as a human being and frees himself from the slavery of matter. After giving thanks, we move on to the recollection phase, remembering past events when God has saved and then calling upon the Holy Spirit for help. The task of the Holy Spirit is to cleanse the water from evil forces and sanctify it.

Consecration, blessing of water, does not appear physically. According to the prevailing models and objective laws of this world, nothing really "happens" to water. According to Schmemann, it cannot be demonstrated by laboratory experiments. He writes that in consecration, there is no attempt to change or replace water with something supernatural, but to restore it to its original state and connection with Christ. Expecting such changes has always been considered blasphemy or sin by the church.

The consecration of water is, therefore, a recreation of matter and the world, and it is given to man as a gift. Therefore, the blessing of water makes man divine. In addition to Orthodox blessing of water, this ceremony is also performed in other religions, such as Catholicism, where it is also common to sprinkle this water in different places to bless them. Catholics often dip their fingers into holy water and make the sign of the cross when entering the church, thus reminding them of the baptism they have received.

I'm left pondering what I read, hoping that Schmemann's view is just an extreme and that reality is different. I reflect on the water studies I described earlier. The studies are based on the idea that something can change water, and the change could also be verified. So is doing research sin and blasphemy? Perhaps such thoughts were also the reason why the church and science diverged already in the 17th century. In extreme cases, the church even considers scientific work as sin. After the separation, it was possible within the scientific community to study various natural phenomena more openly without the church dictating what can be studied. Now the direction of research is no longer dictated by the church, but it is still controlled and guided in many other ways. Although there is a lot of discussion going on about open science and research, in practice, the direction is strongly influenced, among other things, by worldview and the established models we have. However, I don't want to delve deeper into all of this here. I strive to adhere to the principles and fundamental values I raised in *Hyperreality*. The most important thing in my own actions is justice, ethics, freedom, and the innate human desire to know and understand.

The church may not, thus, really have an interest to show or even take a stand on whether blessed water is different from unblessed water. It is not essential to them. The Bible also writes about healing water, such as the Jordan River, which connects the Dead Sea and the Mediterranean Sea. The Jordan River is a holy river for both Jews and Christians. The Jordan River was a river that mysteriously divided so that Joshua could lead his people across it into the promised land (Joshua 3–4). It was the river where John the

Baptist baptized Jesus (Matthew 3:13–16). In the books of Kings, there is also the story, among others, of the commander Naaman, who bathed seven times in the Jordan River and was cured of leprosy (2 Kings 5:10).

It seems that within the church, it is thought to be pointless to conduct experiments with blessed water and find out if it is different from unblessed water. However, holy water is available in various parts of the world from sacred places, and this water has been studied.

SACRED PLACES AND THEIR WATER

Spiritual traditions around the globe revere water as sacred, using it in purification and blessing rituals and ceremonies that connect them to the source of all creation.

There are many sacred sites where water is considered particularly healing or therapeutic, such as Lourdes in France.

Bernadette and the Grotto of Lourdes

In February 1858, 14-year-old Bernadette Soubirous experienced an apparition of the Virgin Mary. For 15 consecutive days, the Virgin Mary appeared and guided Bernadette. Among other things, she directed Bernadette to a nearby cave near Lourdes (Grotte de Massabielle) to find a spring, for from the depths of this spring's heart, one can reach the very life of God, the source itself.

In that spring, Bernadette experienced a total of 18 apparitions that the Catholic Church recognized as miracles several years later. Bernadette lived the rest of her life in a convent. In 1933, 50 years after her death, she was declared Saint Bernadette, the patron saint of the sick and poor.

The Grotto of Lourdes remains a popular pilgrimage destination, visited annually by millions of pilgrims. People drink the water flowing in the cave and use it to wash their faces and hands, believing in its healing properties. The allure of Lourdes is further enhanced by reports of miraculous healings from various illnesses and ailments said to have occurred there. After

Bernadette's apparitions, over 7,000 miraculous healings are reported to have taken place.

One such case was that of Louis Bouriette, who had lost his sight in an explosion accident 20 years earlier. Bouriette was skeptical, but at his sister's insistence, he agreed to visit the cave. There, he rubbed the muddy water from the spring into his eyes and immediately regained his sight.

The Catholic Church established its own group to investigate these reported miraculous healings, and nowadays, each report goes through a three-tier process, classifying it as a "reported," "unexpected," and ultimately "confirmed" healing. Of the cases presented, the church has accepted 67 miraculous healings (as of 2022). In recent years, few healings have been documented, as in 2008, due to medical opposition to miraculous healings, only healings significant by medical standards, which cannot be explained without Lourdes or its healing water, have been recorded.

Lourdes water has also been studied. In a study by A. Ansaloni, it was found, among other things, that if ordinary water, initially containing free chlorine, was also supplemented with Lourdes water, it retained the chlorine content better than just plain water. Furthermore, the decrease in chlorine content was independent of the ratio of Lourdes water added, whether it was 1:200 or very little (1:1 x 1023), as if only knowing the water being added to the sample was enough. When studying the structure of Lourdes water, it has been found to be highly organized (EZ), meaning it has a regular structure that also remains stable.

Often, water from sacred places is kept as a talisman. The talisman can be made of metal, glass, or other materials suitable for water, with the purpose of preserving the water's energy near its bearer. When I was around eight or nine years old, I carried a tiny glass bottle with me everywhere, containing Cologne water (Eau de Cologne 4711). It wasn't holy water, of course, but it was my treasure. While I was playing, it contained power water, the drinking of which gave my dolls superpowers. Cologne water is much more than just water. Its exact recipe is not disclosed, but it contains at least lemon, bergamot, orange, rosemary, lavender, and ethanol. It's not actually suitable for a child, but it never occurred to me to drink it. I was more enchanted by its scent, and I can still recall it right now.

Nowadays, I often wear a water talisman on my wrist as well. It's a spiral bracelet with a glass sphere containing water from Damanhur spiritual community, Piedmont Prealps, Italy. It is said that the bracelet balances, aids in inner dialogue, and connects the wearer's different aspects into a harmonious whole, so they can fulfill the task they came to this life to fulfill. When wearing the bracelet, I feel that the most essential part of it is precisely that glass sphere.

Sacred places have been revered throughout history, and they exist all over the world. Water is often intimately linked to the sacredness of these places, as ancient cultures considered water central to their ceremonies and rituals. Water was a sacred element. For example, as Christianity gained ground in ancient Rome, churches were often built on the sites of baths because water was already available there for the church's needs as holy water. Baptism has always been and continues to be significant for Christians. In addition to Lourdes, the Jordan River, and the Ganges, other sacred water sites include:

- The sacred lake of Karnak in Egypt. The lake was built by Pharaoh Thutmose III (1473–1458 BCE) and is essentially a basin measuring 120 m x 77 m (394 ft x 253 ft), which priests used in their rituals. The lake symbolized the source of life.
- Lake Titicaca on the border of Peru and Bolivia is an important place for the Incas. According to ancient Inca mythology, the god Viracocha, "Foam of the Sea", emerged from the lake and created the sun, the stars, and the first humans. The lake is still considered sacred, and pilgrimages are made there, though the water is now quite polluted.
- Cenote Sagrado or the Sacred Cenote in the Yucatán Peninsula at Chichen Itza. The cenote is a 60 m wide pit in the ground with water about 30 meters (100 feet) below, surrounded by limestone walls, although the water level varies. The ancient Mayans sacrificed offerings to the rain god Chaac at the Cenote, sending him valuable stones, gold, and other objects, and even people to appease the god.
- Lumbini in Nepal, the birthplace of the Buddha, and the sacred lake of Maya Devi. Nepal also has Gosainkunda, a sacred lake where, according to

legend, the god Shiva swallowed the poison meant to destroy the world and thus saved the world. Shiva drank water coming from the mountains and was saved. Pilgrims on their journey dip into the icy water, repeating Shiva's name.

• The wells and tower of Glastonbury, England. A sacred place for both Christians and ancient Druids, where, according to one theory, the Holy Grail is buried.

These sacred water sites are just a few examples, but there are numerous sacred water sites and springs in almost every country. People have traveled, and still do, to these places from afar. One such place is Baden-Baden in Germany. "Baden" means bath, and the place was founded during Roman times. The area has 29 natural hot springs, with water temperatures ranging from about 122–158 °F (50–70 °C), and people came there to treat various ailments.

The Italian word *"pellegrino"* means pilgrim, and many are surely familiar with bottled San Pellegrino water. San Pellegrino (in Lombardy) has been a pilgrimage destination for hundreds of years, and Leonardo da Vinci is said to have visited the waters of the area. Da Vinci hoped to find answers to the mystery of creation by studying the behavior of water in the earth and air.

I am from Finland which is often referred to as the 'Land of a Thousand Lakes,' boasting over 188,000 lakes throughout the country. In the book *Pyhien vesien äärellä* (in English *By the Holy Waters)*, Tuomo Kesäläinen and Aimo Kejonen list a total of 34 Finnish sacred water sites: springs, lakes, marshes, groves. The list probably does not include all of them, as there are also dozens of places in Finland with names like Pyhäjärvi (Holy Lake), Pyhävesi (Holy Water), and Pyhäjoki (Holy River).

Water has been and still is part of many spiritual traditions and purification, transition, and blessing rituals. It often symbolizes renewal, purification, change, and fertility. Springs have been seen as a gateway to the spirit world or as the eye of God. I will return to the symbols of water later in this book.

Egyptian water ceremony for children

It is said that in ancient Egypt, newborns were bathed in the Nile in a water ritual when they were a week old. The child's family and village elders would walk to the banks of the Nile, singing melodies that thanked the child's soul and asked the gods of the Nile to protect and bless the child's incarnation. Women brought food with them, which was offered as a sacrifice to the Nile, and a miniature boat with a white candle was sent down the river, symbolizing the child's earthly longevity. In many parts of West Africa, this seventh-day ceremony is still practiced; for example, among the Akan and Ga tribes in Ghana, a ritual is performed to ensure that the child does not return immediately to the spirit world.

Water baptism is an integral part of many religions. In the Bible (Numbers 8:7), it is said: *"To purify them, do this: Sprinkle the water of cleansing on them; then have them shave their whole bodies and wash their clothes. And so they will purify themselves."*

When a person is being baptized, it is said of God: "I*n the beginning, you created the heavens and the earth, and your life-giving spirit moved over the waters. Water is your gift, which nourishes and sustains creation. In the Flood, you drowned sin but saved Noah and his family. Thus, you gave us an example of baptism. You led your people across the Red Sea and freed them from Pharaoh's slavery. According to your will, John baptized your Son Jesus in the waters of the Jordan, and you anointed him with the Holy Spirit. On the cross, he shed blood and water from his side, and after rising from the dead, he himself instituted the holy sacrament of baptism for his disciples."* Baptismal water is thought to be a symbolic protection and refuge, and a sign that God never forsakes the person baptized into His care under any circumstances.

Like many other Finnish children, I was baptized around the age of two months. In the Evangelical Lutheran Church, the church ceremonies associated with life's most important stages are baptism, confirmation, marriage, and funeral services. Naming is also an essential part of infant baptism. Actually, infant baptism is often a baby baptism. According to Lutheran theology, a child is born sinful, and in baptism, they are brought from sin into holiness, or into communion with God. In baptism, the child receives faith. Baptizing children or babies is also justified by the words of Jesus: *"Let the children come to me."* In baptism, the child is made a member of the community.

In the Western world, water is considered holy if it has been blessed by a religious or sacred person or taken from a place or source considered sacred. According to the Indian Vedas, however, all water is sacred and fundamental. For example, all rivers are considered holy in Hinduism, especially the Ganges, as it is believed to be the physical embodiment of the goddess Ganga. The Ganges cleanses karma and even residues from past lives. Because water is considered fundamental, it is still common in India to bathe in holy rivers and to immerse various objects in them, including statues of gods, idols. For this reason, people are often cremated, and the ashes are submerged in holy waters, symbolizing purification and return to the source of all.

In his book *Science and Spiritual Practices*, Rupert Sheldrake writes about sacred places, rituals, and connection to nature. Pilgrimages to sacred places have long been a part of spiritual practices. Sheldrake mentions, among other things, Hindu pilgrimages to holy rivers, wells, and especially the Ganges. Pilgrimages are often made on foot. Upon reaching the destination, the sacred place is circled clockwise, and thanks are given. In recent years, making pilgrimages has become more common in Western countries as well; often, visits to sacred places are included in tourist trips. Pilgrimage also symbolizes a spiritual journey that we undertake in this life. Our entire life can be seen as a spiritual journey.

Does holy or treated water heal? Naturally, no absolute answer can be given when speaking of any method related to healing. In theory, one could think that all those ailments and illnesses that the body can heal on its own can be reinforced by water treated with intention. Drinking water in safe amounts has no harm. Of course, this does not mean that water helps everyone.

Dr. Batmanghelidj and healing powers of water

There are, however, many stories about how water has helped a person. Dr. Batmanghelidj wrote an entire book about his experiences. When the Iranian Revolution broke out in 1979, he was sentenced as a to two years and seven months' imprisonment as a political prisoner. It was there he discovered the healing powers of water. One night, one of his fellow prisoners was suffering from crippling peptic ulcer pain, and he needed to treat him. Since there were no medications at his disposal, Dr. B. gave him two glasses of water. Within eight minutes, his pain disappeared. He gave him instructions to drink two glasses of water every three hours and the man became absolutely pain free for his four remaining months in the prison. Dr. B. continued to treat others in a similar way and successfully treated 3,000 fellow prisoners suffering from stress-induced peptic ulcer disease with water alone. The prison was a perfect place to carry on these studies and he actually chose to stay an extra four months in prison to complete his studies into dehydration and bleeding peptic ulcer disease. An article about his findings was published in *the Journal of Clinical Gastroenterology* in June 1983 and he also wrote a book called Your Body's Many Cries for Water about his findings.

The question of whether holy or treated water heals, however, risks reducing the scope of the inquiry to essential aspects of water. The more one becomes acquainted with water, the more it seems that one essential aspect of its healing effect is precisely the connection and interaction with water. That is, the most essential thing doesn't seem to be focusing on curing symptoms but rather the connection to water on a deeper level. Water cannot be viewed solely as an objective subject of study but must also be seen as part of its researcher. With this in mind, I decided to conduct my own water experiments.

CHAPTER 9. MY OWN WATER EXPERIMENTS

"Water is the driving force of all nature."
– Leonardo da Vinci (1452–1519)

As I was writing this chapter, I met my friend, author and entrepreneur, Maria Nordin. The water of the new well at Maria's summer cottage had been cloudy and smelled like gasoline for a year. One day, while at the cottage, she performed an exercise she called lake meditation, in which she visualized a connection to the water by imagining its clarity in her mind. The next morning, the well water at the cottage was completely clear and odorless. Maria was sure it was also suitable for drinking. The water was also tested and found to be completely clean and suitable for drinking.

Earlier in this book, I discussed studies related to influencing water. In *From Quantum Physics to Energy Healing*, I did some preliminary measurements on how energy healing and intention affect the pH level or acidity of water. The idea behind energy healing is that it affects the recipient in the best possible way, and one might expect the pH to move towards the optimal pH for humans, which is said to be 7.30–7.45, which is also the pH of human saliva.

In fact, this is exactly what happened. The results of my pH measurements showed a trend towards a neutral value and also towards the optimal value for humans. Another series of measurements conducted as an experiment yielded similar results. However, these are only two series of measurements, and it is not possible to draw final conclusions based on them alone, as further measurements would be needed.

I decided to continue my experiments with water. How would the results differ when treating water for 15 minutes and then another 15 minutes? I had filtered tap water, spring water from Hardanger, Norway, and Grander water. I measured the pH of these using a pH meter with a precision of 0.01. I also have to mention that the tap water I used was from Helsinki and it's drinkable and said to be the world's best tap water.

The spring water from Norway was branded Isklar, and its mineral contents were as follows (mg/l): Na 1.1, Ca 7.9, Mg 0.4, K 1.9, Sulfates 6.5, Bicarbonates 16.2, Chlorides 1.9, Si 1.2.

Grander water is the life's work of Johann Grander. Grander understood the importance of naturalness early on. By observing nature, he concluded that water has the ability to transmit information. Grander water is a water revitalization method developed by Grander, which is currently used in many spas, swimming pools, and hotels across Europe, especially in Germany. His goal was to transfer the properties of vitalized water to all waters accessible to humans, animals, and plants. The unit used for water revitalization does not require an external power source, electricity, or chemicals, and it does not require maintenance. The units are made of steel, and their core element is water, treated according to Grander's method to become "extremely coherent, functional water." Grander water and the water being treated are not in direct contact within the unit. Typically, the unit is installed on a water tap or elsewhere in the water pipeline.

Naturally, as Grander is a commercial entity, precise information about the technology and structure of Grander units is not provided. However, there could be some technology generating a weak magnetic field in the unit. They refer to a study by Semmer et al., that examined calcium carbonate nanoparticles referred to in the publication as DOLLOP (*dynamically ordered liquid like oxyanion polymers*). These particles are also present in ordinary tap water. DOLLOP particles can arrange themselves in new ways under certain conditions, including the influence of a magnetic field, even a weak one. Water itself is a diamagnetic substance, meaning it is not attracted to a magnetic field. In other words, Grander water may contain impurities from the water's perspective, mineral particles that make the water "better." The publication suggests that this could also be an explanation for water memory.

Thus, I decided to measure the pH of these waters. I first performed energy healing on the water for 15 minutes, and then another 15 minutes, and measured the pH after these sessions. Additionally, I measured the pH 3 hours after the end of the energy healing session. I included water directly

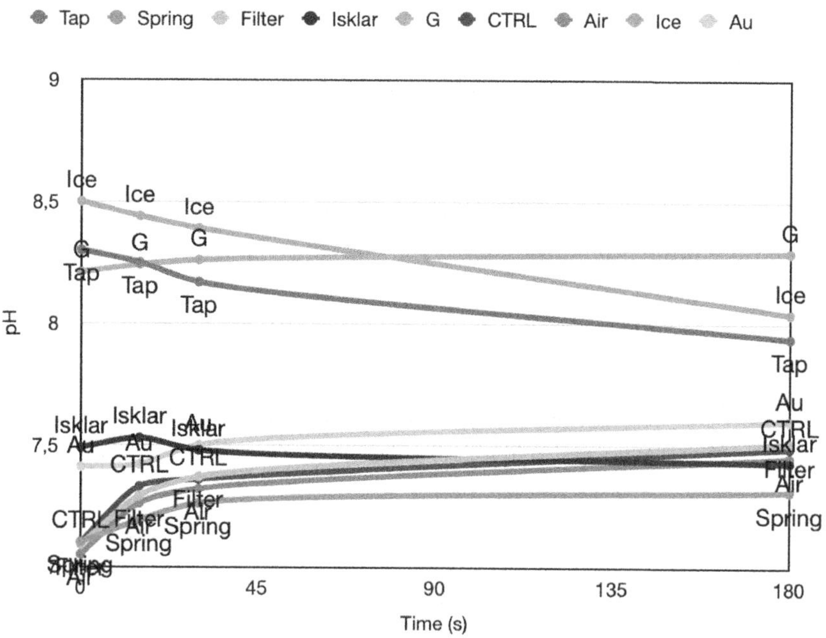

Figure 17. pH of different water samples as a function of time.

from the tap (Helsinki water) in the measurements. This is because Grander water is mentioned in connection with unessential contact with the water it affects. All waters were stored in the same refrigerator before measurements, so I could perhaps rule out any potential effect of Grander water in this way.

In addition to the above, I decided to include filtered tap water on my list, which was simply transferred from one glass to another and shaken to oxygenate the water. Bengston, among others, performed this in his water publication, which I mentioned earlier in the book, noting that this process, resembling homeopathy, seems to have an effect on water. However, Bengston also diluted the treated water before shaking it. Yogi Ramacharaka also mentions in thc book *Water Cure: The Hindu-Yogi System of Practical Water Cure* the "pranaization" of water: *"In order to Pranaize their drinking water, the Hindus pass it through the air from one vessel to another, thus imparting to the fluid a new 'life' which is quite noticeable, and which may be proven by anyone who will take the trouble to experiment with it. Water thoroughly Pranaized acts as a decided stimulant and invigorant, and imparts new vitality and energy when used in normal quantities. Who does not know the invigorating effect of a cupful of clear, vitalized spring water, or water from a mountain stream? There is a world of difference between such water and the lifeless water from the city water mains. And yet the latter may be Pranaized or vitalized by the exercise of a little trouble and work, in the direction of passing the water through the air in the manner just mentioned."*

The measurements included water taken directly from the tap, filtered water (using Brita MAXTRA+ water filter cartridge), bottled Finnish Spring spring water, Norwegian Isklar spring water, thawed (filtered) water from ice in the freezer, filtered tap water left in direct sunlight for at least an hour outdoors, and filtered tap water oxygenated (pranaized) before and after the treatment session by transferring it from an empty glass to another and shaking it. Additionally, a control water was taken for these waters, which was simply kept on a shelf in another location. The control water was filtered tap water. All others were treated, except the control water.

The pH, conductivity, salt content, and temperature were measured for all waters. Conductivity and salt content are related: the more saline the water, the more conductive it is. Conductivity depends on the temperature of the water. The pH of water at 77 °F (25 °C) should be around 7.

Additionally, I decided to investigate the variation in the magnetic field during treatment and measurements. Changes in the magnetic field in the vicinity of healers have been mentioned in some studies, including Bengston and Moga, who refer changes occurring near healers, but not near the treated individuals.

For measurements, I used the Water Quality tester device (EZ9909SP, INF Company, Höganäs, Sweden), which measures pH, conductivity, salt content, and temperature. The resolution of pH was 0.01 units, conductivity 1 μS/cm, salt content 1ppm, and temperature 0.1°C.

pH

I aimed to measure pH values close to the temperature of 77 °F (25 °C) to prevent temperature changes from affecting the pH readings. However, in practice, the temperatures slightly decreased, especially in the results after three hours, which should be taken into account when interpreting the results. The pH meter was calibrated with a buffer solution with a pH value close to the sample's pH value, that is water with a pH value of 6.86 at 77 °F. Between each measurement, I rinsed the pH meter electrode with tap water and also with the water sample being measured several times before determining the sample's pH value.

Water sample (200ml)	Abbreviation in the image	pH, conductivity, salinity and temperature in the beginning	15 min after the energy healing	30 min after the energy healing	3h after the end of energy healing session
Tap water	Tap(A)	8,30: 162μS/cm: 80ppm: 25,3°C	8,25: 159: 79: 24,1	8,17: 157: 77: 23,9	7,94: 153: 74: 20,4
Bottled spring water	Spring	7,10: 42: 21: 25,3	7,19: 43: 21: 24;0	7,26: 43: 21: 23,1	7,31: 42: 21: 20,3
New filtered tap water	Filter	7,09: 146: 73: 26,4	7,29: 150: 75: 24,3	7,37: 149: 74: 23,6	7,51: 154: 77: 20
Spring water from Norway	ISKLAR	7:49: 55: 28: 25,5	7,53: 55: 28: 24	7,48: 58: 29: 23	7,43: 56: 29: 20
Grander water	G	8,21: 384: 192: 8,22	8,24: 380: 190: 23,7	8,26: 388: 194: 22,8	8,29: 391: 195: 19,8
Control water (filtered tap water)	Control.	7,10: 150: 75: 27,8	7,33: 148: 74: 24,4	7,36: 148: 74: 22,9	7,49: 151: 75: 20,9
Filtered tap water, which was transferred from one glass to another (oxidation)	Air	7,05: 170: 85: 28,7	7,26: 156: 78: 24,1	7,32: 162: 81: 23,6	7,45: 163: 82: 20,1
Water melted from ice	B	8,50: 163: 81: 27,2	8,44: 161: 80: 24,8	8,39: 163: 81: 23,7	8,04: 164: 82: 20,3
Filtered tap water, that was kept in sunlight	C	7,41: 150: 74: 26,6	7,42: 149: 74: 24,6	7,50: 149: 74: 23,8	7,60: 151: 75: 20,3

Table 2. pH, conductivity, salinity, and temperature of different water samples at the beginning, after 15 minutes and 30 minutes, and 3 hours after the end of the treatment.

The table below shows the pH values of the measurements, as well as conductivity, salinity, and temperature. The graph depicts the pH values over time. Values in the table are given in units measured by the Water Quality Tester.

The most significant changes during the 30-minute treatment occurred in oxygenated water and filtered water (0.27), control water (0.26), bottled spring water (0.16), and tap water (0.13 units). When examining the measurements after 3 hours of treatment, the most significant changes occurred in thawed water (0.46), followed by filtered water (0.42), oxygenated water (0.40), control water (0.39), tap water (0.36), and bottled spring water (0.21). Among these, all values increased except for thawed water, which decreased by 0.11 units during the 30-minute treatment and by as much as 0.46 units after 3 hours! Similarly, the pH reading of tap water decreased, but to a slightly lesser extent (0.36 units). The pH of Norwegian Isklar water also decreased but only very slightly.

The range of pH for good, potable water is quite broad, from 6.5 to even 9.5, so all of these waters are potable based on pH. Typically, water in pipelines is preferably kept at a relatively high pH rather than low to prevent the pipes from corroding due to acidic water.

Temperature also affects pH. As temperature decreases, pH usually increases. For example, the buffer solution used to calibrate the pH meter at the beginning of the measurement had a pH of 6.86 at 77 °F (25 °C), but at 68 °F (20 °C), it was 6.88 (a change of 0.02 units). Changes greater than this should be considered significant.

What do these pH measurements tell us? Firstly, it is peculiar that the pH of thawed water and tap water only decreases after 3 hours, even though the temperature has dropped by 41–44.6 °F (5–7 °C) since the beginning of the measurement. Secondly, you probably noticed the changes occurring in the control water as well. Is it possible to distinguish between the waters, or is there a connection between these samples? What effect does it have that I always rinsed the pH meter with tap water before rinsing it with each sample's water? Is the change in pH truly due to the energy healing? Is it even possible to observe the effect of energy healing on the pH of the samples?

The change in pH seems to continue even after 30 minutes of treatment. However, this is not unusual for energy healing, which is a non-local and time-independent phenomenon. It is also essential to consider what previous water studies have shown: The effect may not always be immediately apparent when the studies begin but may occur even before the start of the experiment.

In my book *From Quantum Physics to Energy Healing*, the most significant change in pH measurements occurred in tap water, which dropped by 0.5 units over 2 hours. Similarly, in this measurement, the change in tap water was almost the same, 0.36 units. The pH appears to approach the optimal pH for humans, but the minerals in the water surely also have an effect on the water, its behavior, and properties. These are also relevant to humans. Drinking completely pure, distilled water is not healthy due to mineral deficiencies. The body's fluid and salt balance are disrupted, and minerals and salts are lost from the body through increased urine excretion. The differences in the numerical values between this experiment and the previous ones may also partly be due to differences in the measurement devices used, accuracy, and of course, differences in the water itself, as waters are not consistently uniform but also vary.

The pH of Grander water and thawed water is higher than that of the others, as one might expect. Thawed water should contain more EZ water. However, in this case, the waters were heated to the same temperature. Perhaps the organization of the thawed water has not had time to dissipate yet? Grander water is also advertised as particularly healthy and mineral-rich, so its pH could be expected to be more alkaline. The higher pH of tap water is likely due to chemicals added to the water for pipeline maintenance. Unfortunately, with my home equipment, I could not measure which chemicals were present in the water, but certainly chlorine is added to water. Fluoride is not added to water in Finland, but it may naturally occur in small amounts.

CONDUCTIVITY

Water conductivity increases as the ion concentration rises. Pure water itself is not a good conductor of electricity; rather, water is an insulator. Compounds with pH close to neutral also have lower conductivity. In addition to ion concentration and temperature, the type of dissolved substances in the water also affects conductivity.

Typically, the conductivity of drinking water ranges from 0.005 to 0.05 S/m (= 50-500 μS/cm). Very pure water can be as low as 5.5×10^{-6} S/m (= 0.055 μS/cm), while seawater is approximately 5 S/m (= 50,000 μS/cm). The higher the reading, the more impurities are present in the water. Thus, the reading indicates water quality in addition to conductivity.

Among the water samples, bottled spring water from Finland (42–43 μS/cm) and water from Hardanger, Norway (55–58 μS/cm), had lower conductivity, meaning they contain fewer impurities than others. On the other hand, Grander water (380–391 μS/cm) was significantly more conductive than others, with the conductivity of other waters ranging from 146–170 μS/cm. The conductivity of Grander water is likely due to minerals or chemicals present, which make it not only more conductive but also more alkaline.

During energy healing, conductivity in the samples remains nearly the same or slightly increases, except for filtered tap water transferred from one glass to another, which decreases from 170 μS/cm to 156 during the 30-minute treatment, ultimately ending at 162 μS/cm. Perhaps it is due to mixing and shaking, that negative ions "escape" from the water into the air or onto the edges of the containers, resulting in a decrease in conductivity?

SALINITY

Saline water contains, as its name says, salts, with the most common being ordinary table salt, sodium chloride NaCl. The concentration is usually expressed in units of ppm (parts per million). For example, if the concentration of dissolved salts in water is 10,000 ppm, then 1% (10,000 divided by 1,000,000) of the water's weight comes from salts. The salinity of pure water is typically below 1000 ppm, whereas seawater is around 35,000 ppm.

In the measurements I conducted, salinity largely follows the same patterns as conductivity, with a maximum of 195 ppm in Grander water. During energy healing, there are hardly any changes in the numerical values.

VARIATIONS IN THE MAGNETIC FIELD

Variations in the magnetic field during energy healing have been measured in publications such as Bengston and Moga's study. Their research found changes, especially in the vicinity of the healer, rather than in the person receiving the treatment. Changes were observed both in close proximity and at a distance during treatment. For the duration of treating the waters, I also placed a phyphox mobile app magnetometer to the table to measure changes in the magnetic field around me. The magnetometer is very sensitive to various movements, so it was placed a bit further away from the samples on another table to avoid measuring minor movements.

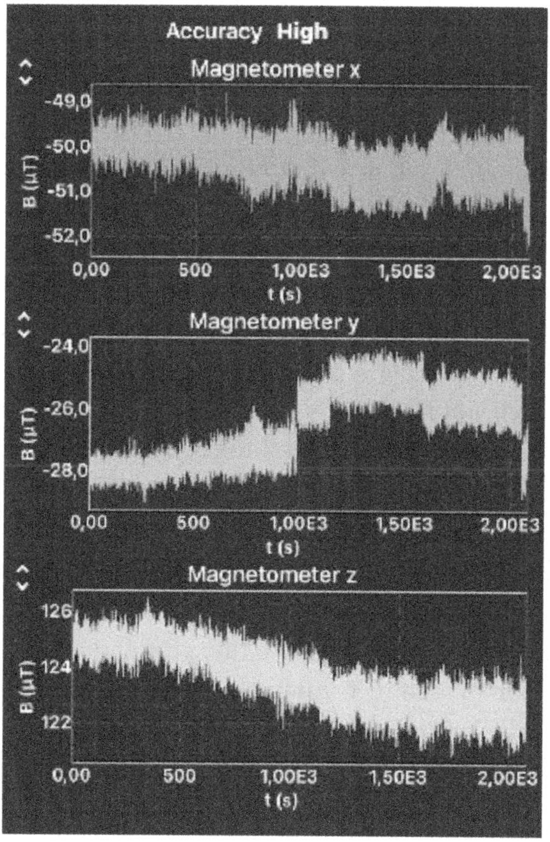

Figure 18. Magnetic field variation measured by the magnetometer.

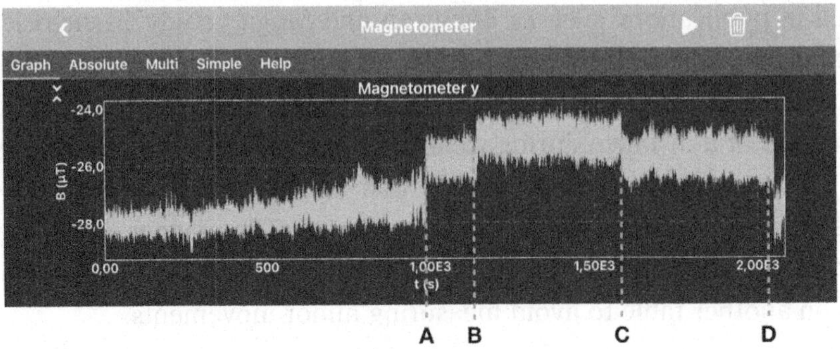

Figure 19. In Figure A - Start of treatment (1000 s), B - Connection to the treated water (1200 s), C - Connection disappears and the sense of treatment necessity (1600 s), D -Treatment ends (1900 s).

The accompanying figures (Figure 18 and 19) show data measured by the magnetometer on the x-, y-, and z-axes. Each graph shows the magnetic field B (in μT) as a function of time t (in seconds). I started the measurement while taking initial readings (pH, conductivity, salinity) before the first 15-minute series. The actual measurement (first 15-minute treatment series) started at point 1000 s (A), where a clear step is visible in the y-data (Figure 19). Another step is seen approximately three minutes into the treatment, at around 1200 s (B), when I felt a connection to the sample waters. It was easy for me to visualize them, feeling as if I could see the water's "journey," everything it had "experienced," through all these samples. Around 1600 s (C), that feeling disappeared, and I felt there was no need for further treatment: What was meant to happen had already happened. The first treatment series ended when the time in the graph reached 1900 s (D). However, I treated for a full 15 minutes before taking water quality measurements, after which I treated for another 15 minutes. In the graph of the latter 15-minute series, there were no similar clear steps visible. Nevertheless, as the healer, I felt that the treatment was already underway as I continued.

Points B and C in the graph are particularly interesting. The experience of establishing a connection to the treated water is evident at point B, and the sensation of the connection disappearing at point C. The magnetometer is very sensitive and was positioned at a sufficient distance from the treatment point precisely to avoid being moved. It is possible that some physical movement occurred at the start and end of the treatment, which the measurement could pick up, but at points B and C, I did not observe any such movement or other changes; I only experienced the sensations of connection to the water in my mind.

GROUP INFLUENCE

Lynne McTaggart's experiments on group influence inspired me to conduct my own experiment with students studying to become Reiki teachers. During one teaching session, I brought bottled spring water for us to treat. Another bottle of spring water I had with me was not treated but remained in my backpack throughout. At home, I had a third bottle of spring water, which I placed on a shelf in my workspace as a control. The Reiki students were unaware of these two other bottles of water.

During our gathering, we sat in a circle, meditated first, sent remote healing, and also treated the water samples. The water treatment lasted about 10 minutes.

Considering that the effects of the treatment could potentially be enduring, I conducted the measurements three weeks after the meeting. Would the measurements reveal any differences in the waters? The differences might be small, but I decided to try nonetheless.

We repeated the treatment later when we met again in the same manner, although this time the healers were aware of the other two bottles of spring water as well.

The setup for my experiment in the measurements was as follows:

1. I poured 100 ml (3.38 ounces) of water into three beakers. A was the treated water, B was the reference water that had been with me in my backpack, and C was completely separate from these. I measured the pH, conductivity (μS/cm), salinity (ppm), and temperature (°C) of the samples, and additionally the surface tension (mm) using a capillary tube. The results of these measurements are found in the table below. I did this both after the first and the second treatment sessions.

2. With the latter treated water, I also conducted this experiment: I placed organic alfalfa seeds in three glass jars. Alfalfa seeds are very small, and I measured approximately 1 teaspoon of them into each jar. The jars were also labeled A, B, and C, and I watered the seeds with the corresponding waters. Would I observe any differences in germination and growth between the waters this way?

I conducted the measurements the same way for all: I calibrated my meter before measurements and between measurements in a buffer solution with a pH of 6.86. Before each actual measurement, I rinsed the measurement head with water from the respective bottle a couple of times. The results of the measurements are in the table below, with the second series always in parentheses after the first.

	A	B	C
pH	6,61 (6,64)	6,44 (6,54)	6,44 (6,48)
Conductivity (µS/cm)	48 (47)	48 (45)	48 (47)
Salinity (ppm)	24 (0,0%) 23	24 (0,0%) 23	24 (0,0 %) 24
T (°C)	19,6 (19,1)	19,4 (19,5)	19,5 (19,6)
Surface tension (mm)	35 (37)	32 (33)	30 (27)

Table 3. pH, conductivity, salinity, temperature, and surface tension of bottles A, B, and C. Bottle A was the treated water, B was the reference water with me during the treatment, and C was the completely separate water.

The only differences in the numerical values are practically visible in the pH and surface tension. The pH value of the treated water is the highest, and in the first measurement, B and C are the same. In the second measurement, B falls between A and C. The surface tension of A is the highest; B and C are again nearly the same in the first measurement, and in the second, B is higher than C.

Although the differences are small, I am surprised that they are in the direction one would expect. The second measurement also surprises: now that all the healers were aware of bottles B and C, could the higher reading of B be explained by this?

If treating the water transforms it into a better form for humans, one could assume that it becomes more organized, and therefore, the surface tension would also be expected to increase. Perhaps that's why the surface tension of A is the highest.

ALFALFA AND THE TREATED WATER EXPERIMENTS

In the second phase of the experiment, I sprouted alfalfa seeds in three different waters. Alfalfa seeds are very small, and I measured about one teaspoon of them into three different containers. The seeds had been soaked overnight first. I used each water sample for soaking. The next day, I poured out the soaking water and left the seeds to sprout. The packet of alfalfa seeds says the sprouting time is 3–4 days. Below, you'll find pictures of the jars on each day.

Some of the seeds in Dish A seemed to have already sprouted when I poured out the alfalfa-sprouting water. There were about 40 sprouted seeds in Dish A, while in Dish B, there were about 20, and in Dish C, about 15. The seeds in Dish A appeared to be growing faster than those in the other two dishes. On the second day, Dish A had the most long white sprouts, while Dishes B and C had roughly the same amount. Some of the seeds in Dish B grew very strongly, while others grew more slowly. The difference between the dishes seemed to be greatest on the second day, but by the third day, it seemed to have leveled out.

This is, of course, a single set of measurements, but the result is consistent with studies investigating the effects of energy healing on plants, such as those of Gradin and others. I reached a similar conclusion when growing watercress seeds in my book *From Quantum Physics to Energy Healing.*

In some energy healing studies, there has been discussion about what duration of energy healing is sufficient, and whether it's even possible to determine the necessary dose. This time, we treated the samples for only 10 minutes. Would the result have changed if we had repeated the treatment again? On the other hand, if the change is enduring even after a short interaction, a new session might induce new changes only if something has changed or strengthened. However, it might still be interesting to investigate

Figure 20 a, b, and c. ALFALFA dishes on days 1, 2, and 3. The top is treated water A, bottom left is B, and bottom right is C.

whether the ratios change if the water is treated for longer. Also, how dependent are the results on the healer? In measurements investigating magnetic field variations, for example, the readings near experienced qigong masters were multiple times higher than those of other energy healers.

WHAT DO THESE RESULTS TELL US?

The impact of intention may not necessarily be evident in the water quality parameters that I examined in the experiments of this chapter. Pyatnitsky and Fonkin found in their study that human consciousness seems to affect the structure of water, and the change was more apparent in the structure than in the properties of water, which are, however, dependent on the structure, such as water conductivity and surface tension. Some variation between different waters was observed between both different sample waters and water treated by the group and other waters.

Radin et al. previously found in their water studies that changes were seen only in distilled water, not in bottled Fiji water. This may indicate that the minerals in water have a significant impact on it. This could also explain why, for example, the most changes in pH occurred in tap water, in water melted from ice, and in filtered, "pranaized," and control water. All of these likely have more impurities than bottled waters and Grander water.

Although the differences in measurements were small, the effect of water on plant germination is actually astonishing: seeds sprouted with treated water did initially grow faster than others! After the measurements, I noted that, as in the studies conducted by Radin et al. and in the studies described in my *From Quantum Physics to Energy Healing* book, in this experiment, all seeds watered with treated water sprouted. So far, all seeds I have watered with treated water have sprouted, which is very rare.

Although the differences are small, it's interesting that they are consistent. It would be intriguing to conduct a more thorough study, in which all possible sources of error could be ruled out. It would also be interesting to clarify the structure of treated water and any differences from untreated water, as well as the effect on seawater: what changes occur in it due to treatment? However, water is a challenging subject to study because it is difficult to differentiate between different samples. Montagnier already mentioned this in his own research, where information seemed to transfer non-locally between two samples. Perhaps that's also the reason why, when observing the growth of alfalfa seeds, the samples begin to resemble each other? When documenting the growth of seeds, I ensured that the containers were treated the same on measurement days and brought them close to each other.

Perhaps the same phenomenon occurs with them, meaning that information is transferred between them?

Of the experiments in this chapter, I was most surprised by the difference in the growth of alfalfa seeds, but also by the change in the magnetic field when the connection is made during treatment. The difference observed in pH and surface tension between the group-treated water and the control waters was also surprising, albeit expected. It indicates a connection between consciousness and water and demonstrates that this link can be made visible even through home measurements. However, water must be studied as an anomalous compound in a different way than we are accustomed to, understanding that our thoughts and emotions have an impact. Water research can open – and has already partly opened – a truly revolutionary perspective on our reality.

CHAPTER 10. THE NEXT STEP OF SCIENCE

"Science cannot solve the ultimate mystery of nature.
And that is because, in the last analysis,
we ourselves are part of nature and
therefore part of the mystery
that we are trying to solve."
– Max Planck, Where Is Science Going? (1932)

~

At this point, I need to pause writing for a while and think. I have studied dozens, perhaps hundreds of water studies and have also conducted some of my own preliminary studies. They seem to tell a story about water very differently from the one we have previously learned. Individually, single experiments may not be very significant, but together they tell the same story: We are not separate from water and we can influence it with our thoughts.

AS SIGNIFICANT A DISCOVERY AS QUANTUM THEORY

In the chapter addressing water memory, I mentioned Luc Montagnier, a French virologist and winner of 20 major scientific awards. Among other honors, Montagnier received the Nobel Prize in Medicine in 2008 with Françoise Barré-Sinoussi, for the discovery of the human immunodeficiency virus (HIV). In 2009, Montagnier published two studies on electromagnetic waves originating from highly diluted solutions of various pathogens' DNA. According to Montagnier, these signals could reveal the bacterial or viral origin of diseases such as autism, multiple sclerosis, Parkinson's disease, and Alzheimer's disease. Theoretical chemist Jeff Reimers from the University of Sydney stated in New Scientist, in January 2011: *"If the results are correct, these would be the most significant experiments performed in the past 90 years and demanding re-evaluation of the whole conceptual framework of modern chemistry."*

Montagnier is not the only Nobel laureate who, after receiving the Nobel Prize, has become interested in water research and recognized the extraordinary nature of water. Cambridge University Professor Brian Josephson, who received the Nobel Prize in Physics in 1973, believes water is leading a revolution in physics – not unlike the emergence of quantum theory in the early 20th century. Josephson was awarded the Nobel Prize for his work on superconductivity and quantum tunneling phenomena and for his discovery of the "Josephson effect." In the "Josephson effect", when two superconductors are brought close together, a barrier forms between them, an obstacle they cannot penetrate. This is actually a quantum phenomenon that occurs on an everyday scale. It demonstrates that the phenomena of quantum mechanics are not limited to the scale of atoms and particles. Perhaps it is not surprising that in the 1970s, Josephson became interested in Transcendental Meditation and increasingly in pioneering research at the boundaries of science. He has explored the connection between quantum mechanics and consciousness and the mysticism of the East, a kind of quantum mysticism.

Josephson was actually one of the first researchers to hear about Benveniste's research at a conference in the 1970s and persuaded the Cavendish Laboratory, where he was working at the time, to invite Benveniste to lecture on his research. This lecture led to years of collaboration.

According to Professor Josephson, water has significant structural and dynamic properties that were already revealed in the research of Benveniste and Montagnier. Josephson believes it is a shame that the memory of water is still being dismissed with three non-scientific arguments. The first of these contends that if there are no longer any active molecules in the water, there cannot be an effect. But this doesn't account for the possibility that the structure of the water may have already been permanently altered by the influence of the molecules: the molecules have already changed the water. It is also often claimed that hydrogen bonds, when formed and broken rapidly, are so short-lived that they cannot retain a memory for long periods. However, Josephson reminds us that there are different time scales for different things. Duration may not be relevant, but interaction is. The third argument presented is that if water had memory, then of course it would remember everything it has come into contact with previously. However, the same could be said for magnets, and no one considers them extraordinary.

With regard to this last point, it is interesting to note that magnetism was also referred to as "fluidism." Fluidism was based on the theory developed by the German physician Anton Mesmer (1734–1815) in the late 18th century, according to which universal forces or flows – fluids – permeate the human body and the universe. According to Mesmer, a magnetizer could manipulate these flows and cure diseases caused by the imbalance of flows.

Physics was created to explain the material world, but once again, I find myself in the same situation as I have described often before in my books: Classical physics encounters difficulties when trying to explain phenomena in which the mind and consciousness are involved. A new model should therefore be adopted that can describe all of these phenomena. I have addressed these models in this book. It should be possible to explain the organization of water, that is, organized complexity. There are indeed such theories, including David Bohm's theory of implicate order.

According to Bohm, behind the reality we perceive and its organization in the universe lies an implicit order. This order also includes disorder (fluctuations). In physics, disorder is called entropy, and it indicates the amount of disorder in a system. Due to equilibrium, entropy also has a counterpart called *syntropy*. Entropy is thus related to the increase in disorder and, for example, to how heat escapes from a system, as in my water measurements, where the temperature initially was higher, but because the system was not isolated or even closed, the temperature decreased. Syntropy, on the other hand, is the tendency toward organization, concentration, and the formation of structures. Fantappiè already proposed in 1942 that our physical/material world follows the law of entropy, whereas the spiritual/immaterial world is governed by retrocausality. The future can influence the present moment. Life, thus, would follow the law of syntropy. We constantly balance between these two in life: entropy and syntropy.

In *Hyperreality*, I wrote about the interaction between consciousness and matter, and here I have arrived at the same subject. We explain phenomena that involve the observer's subjective influence as if they were still objective, even though quantum physics has taught us that we are a significant part of reality. For example, the studies of the PEAR laboratory show very well how we should also approach water. In PEAR's studies, it was found that the influence of the observer and consciousness could be seen from the micro-level of particles to the macro-level of everyday phenomena. The challenge in making measurements is that to achieve reliable research results, measurements should ideally be made in thousands, if not hundreds of thousands, because individual events are just anomalies and often small in magnitude, but together they form statistically significant evidence. In my own studies, thus, I may only be able to see direction, whereas a larger number of measurements would likely lead to more conclusive results.

SUBJECTIVE SCIENCE AND EXPERIENCES

In the New Scientist on March 30, 2022, there was an interesting article: *"A new place for consciousness in our understanding of the universe," which stated: "In order to understand mysteries such as quantum mechanics or the passage of time, theorists are trying to integrate subjective experience into physics, as a physical component of our world."* In fact, at least since the birth of quantum physics, the significance of subjective experience has been apparent, and gradually even mainstream Western science has begun to acknowledge this. We must learn to understand subjective experiences and the interaction between mind and matter. There is no such thing as completely objective science, as the observer always influences what is observed. We matter.

"Science seeks to find objective truth, but is such a goal even possible? After all, all objective information is based on subjective observation." (Johanna Blomqvist: *Hyperreality*, p. 26)

"As a result of materialism, the understanding of subjective experiences has been almost ignored. Subjective experiences have been considered partly non-materialistic, falling under the domain of parapsychology." (p. 159)

"We have searched for objective reality, but what we should understand is the subjective." (p. 179)

In their 1997 article "Science of the subjective," Robert Jahn and Brenda Dunne of the PEAR laboratory state that subjective experiences have been ignored for decades, resulting in a completely secular, dichotomous paradigm that is inadequate. Subjectivity must be taken into account. Of course, there are challenges, as subjectivity partly affects the accuracy of results, but it also brings much new insight to research. Although the publication is over 25 years old, unfortunately, it is still relevant today.

Our reality is both physical and mental. I am certain that we are moving toward a holistic science where consciousness plays a significant role because we have no alternatives. A materialistic, even egoistic worldview leads toward destruction. We can never truly separate ourselves from the reality we perceive. So why not use subjective experience more and more as a helping tool for acquiring knowledge?

When water, of which we ourselves are made, is the subject of study, we undoubtedly reach a completely new level in research. Inevitably, in order to *truly* understand water and also ourselves, we must examine it through subjective experiences. In the PEAR studies, the significance of emotional connection was highlighted. More important than maintaining intention in experiments was the emotional connection and a positive, playful atmosphere. Water is often perceived as symbolizing emotions. Perhaps emotions and experiences are the key to deeper understanding, both of ourselves and of water?

DOES RESEARCH STILL MATTER?

"We are part of the universe,
where the work is still in progress...
We are tiny patches of the universe
looking at itself and building itself...
We live in a 'Participatory Universe!'"

– John Wheeler

Above, I've described how many researchers have found something special, meaningful, in water, something that is needed to understand our world, our reality, and also ourselves. There are indeed studies on water, including its lesser-known anomalies, that tell an interesting story. I wonder, does research still matter? Does the world change because new studies tell a different story? At what point does science, or rather the understanding of the world – that is, worldview – change?

Studies that openly approach new topics and present a new approach, perhaps even a new worldview, break the mold of prevailing thinking. At the same time, however, thousands of studies are conducted using prevailing methods and thinking. The risk of alternative perspectives is to become marginal and drown in the mainstream.

Furthermore, research has become a tool of power. It has long been used as a means of influence, and alongside genuine research, distorted research – that is, research made or produced from distorted premises – is also conducted and promoted because it supports a particular mindset or interest. Such research aims to influence and steer people's behavior and perhaps even support the activities or politics of certain companies and organizations. In such a world, science and research – even the real, genuine research done out of a desire to truly understand – lose their meaning. Essential knowledge gets lost amid all the meaningless noise. No amount of research on a topic alone will change prevailing thinking; other methods are needed. At this moment, more than publications, a change in thinking and attitude is needed. Therefore, it is important to share experiences and also different perspectives openly, to surrender to seeing reality in a new way.

People do not always dare to present different ideas. For years, I have been a member of the SSE (Society for Scientific Exploration), which is an expert organization for researchers and scientists investigating unusual or unexplained phenomena. SSE is a group of researchers with an open-minded and inquisitive attitude toward the world, with various perspectives. Some view the world through materialist eyes, while others see consciousness as fundamental. The commonality is that everyone is openly expressing their own minds about things and seeking truth, reality, sometimes even debating about them.

~

It is known that water has a structurally organized form resembling ice but is not ice. It is generally accepted that water, which is already a very special compound, also has special properties. Among other things, the influence of the structured form of water seems to extend much further than previously thought. It seems that water as a research topic has begun to open up to many. The properties of water that have previously been dismissed and overlooked are now being examined more openly than before. Probably, for historical reasons, there is still much to be done. The general public's perception of water as a research topic is still trailing behind.

The bias of public perception can be seen, for example, when searching for information on structurally organized water: *"Hexagonal water, also known as gel water, structured water, cluster water, or* H_3O_2 *is a term used in a marketing scam that claims the ability to create a certain configuration of water that is better for the body."* (Wikipedia)

From this Wikipedia text, one gets the impression that structural water is just a marketing hoax. The author of the text does not appreciate the idea, supported by research, that certain forms of water could be better for the body than others. The article also mentions the fourth phase of water, which is said to be unstable, without mentioning the conditions under which the fourth phase is observable.

Adopting new ideas takes time. Surprisingly, we imagine we understand how things are. We think that progress cannot move forward from here or at least not in a better direction. We understand that previously everything was more primitive, and many things were not known, but as science progresses, we learn more. Rarely does it occur to us that maybe something was already known. In some cases, the truth or reality about certain things may even have been lost. In order for us to move forward, prevailing thinking must be questioned, sometimes even the thought that is assumed to be fundamental. As Buckminster Fuller said, *"You never change things by fighting the existing reality. To change something, build a new model that makes the existing model obsolete."* So, there's no point in arguing and fighting against it. Instead, make a proposal for something new! This is precisely what modern exploration is about: daring to question things and approach what is not known. This same principle applies to many issues. The alternative is to resign oneself to the consequence, which is inevitably eternal ignorance.

What can be done, then, to bring authentic research into the public awareness and give it the possibility of opening people's eyes? If you consider information and its original source reliable, share the information and talk about interesting findings and perspectives. If you are unsure about the accuracy of the information, ask those who are more knowledgeable about it and also consider the motives for publishing the information: Who benefits from it or what follows from the disclosure of the information? Trust your own experience, too. Listen to your own intuition and heart.

CHAPTER 11. WHAT IS THE BEST KIND OF A WATER FOR US?

"On and on the rain will fall
Like tears from a star, like tears from a star
On and on the rain will say
How fragile we are, how fragile we are"

– Sting, "Fragile"

~

I wrote in my travel diary:

I sit on the cabin's terrace overlooking the Geiranger Fjord. The sun is still shining, soon it will disappear behind the snow-capped mountain. The landscape is like something out of a fairy tale. There are already cherries on the cherry tree, a blackbird flies to it to sing. Every now and then, a cold mountain stream of fresh mountain water flows from the nearby waterfall. The water of the fjord is greenish and glistens in the sun. The scenery is beautiful, but at the same time somewhat menacing, as if telling me to enjoy this moment now because it won't last. The height of the mountains, the cold wind, the deep turquoise, calm water, warm sunshine, narrow fjord, and at the same time a giant passing ship, all extremes in this landscape together create this delicate, fairytale-like picture, and there is a feeling that one cannot absorb it enough. My heart feels like bursting, and I don't want to move so as not to break this picture, this feeling.

Water is present everywhere here, life revolves around it. Sometimes the road over the mountains to the village is closed if the roads are too slippery, the weather too snowy, or otherwise challenging. Waterfalls can be seen wherever you look, and I hear their roar, which calms, cleanses, and grounds. During the day, we climbed over half a kilometer to one of the waterfalls, one of the most breathtaking ones. We walked over meadows grazed by goats, along increasingly steep paths. Streams gurgled along the way, and when the water ran out of our water bottles, we could replenish them directly from the stream. The highlight of the trip was walking along a narrow rocky ledge behind the waterfall! Here, the water is truly 'alive,' and its energy is contagious! I feel truly alive and refreshed!

~

If I could choose what water to drink, I would definitely choose the water that flows in mountain streams! So, how can we get the best water for our well-being – "living," energized water? The best water is pure, fresh, and structured (EZ) water, just like the waters of mountain streams or natural springs. This is rarely found in cities, but it's possible to influence water in various ways. In many places, like in Finland, tap water is clean and drinkable, but that is not the case in all countries. There are other important aspects to the quality of water, which I will address here next.

EARLY WATER PURIFICATION

Since ancient times, people have understood the importance of water purity. The earliest known written record of water purification comes from Sanskrit texts and Egyptian wall paintings.

In the Sanskrit text called Sus'ruta Samhita, dating back to around 1400 BCE, various methods of water purification are described. Water could be heated with fire, warmed in the sun, or filtered through sand and gravel. The purification process could also involve using the seeds of the "clearing-nut tree" (*Strychnos potatorum*) or the "*Gomedaka*" stone. The seeds of the clearing nut tree are still used today, especially in India and Myanmar, for the treatment of inflammatory stomach and intestinal diseases as well as water purification. "*Gomedaka*" is composed of zirconium silicate, $ZrSiO_4$, a mineral found in igneous and sedimentary rocks, occurring in yellow, brown, red, and even as dyed tetragonal crystals, the transparent forms of which are used as gemstones. Francis Evelyn Place, who studied ancient medicine in Sanskrit, wrote in 1905, in reference to Ousruta Sanghita from around 2000 BCE: *"It is good to keep water in copper vessels, to expose it to sunlight, and filter through charcoal."* The antibacterial properties of copper were known at that time, as were the effects of sunlight on water and the ability of charcoal to absorb harmful substances.

Drawings of water-purifying devices have been found in the tombs of Egyptian pharaohs including Amenophis II and Ramses II (1500–1300 BCE). Impurities were allowed to settle to the bottom of containers, and then the clarified liquid was separated with a siphon into another container for storage.

According to our current known history, about 11,000 years ago, people began practicing agriculture and establishing permanent settlements. Around the time of the construction of Göbekli Tepe, approximately 10,800–9500 BCE, it is believed that people lived mainly by hunting and gathering. The development of agriculture and permanent settlements was highly dependent on water, especially clean water.

The earliest discovered water supply system with drainage in Europe dates back to around 2000 BCE in Bronze Age Minoan Crete. Urbanization and water systems increased around the Mediterranean region between 500 BCE and 500 CE as the population increased by 10–20%. Even before this, around 4000 BCE in Ancient Greece, boiling and filtration were used for water purification.

Alcmaeon was the first Greek physician to note (around 470 BCE) the importance of water quality to human health. Several of Hippocrates' texts also contain brief comments on the significance and effects of water on people. In his work *On Airs, Waters, and Places*, Hippocrates states: *"Whoever wishes to investigate medicine properly, should proceed thus: in the first place to consider the seasons of the year, and what effects each of them produces. We must also consider the qualities of the waters, and the mode in which the inhabitants live, and what are their pursuits, whether they are fond of drinking and eating to excess, and given to indolence, or are fond of exercise and labor, and not given to excess in eating and drinking."*

Ancient Rome is famous for its aqueducts and baths. Certainly, for the wealthy who had access to private baths and toilets, the situation must have been good, unlike for the poorer people who had to use public latrines and, if possible at all, bathe in public baths. Marcus Aurelius (121–180 AD) described the bath water as a disgusting mixture of oil, sweat, dirt, and water. Little attention was paid to hygiene; there was no running water. Water points were often located in kitchens, allowing diseases to spread effectively as hands were not washed either. During the summer heat, cities smelled disgusting, and especially toward the end of summer during the hottest period, gastrointestinal diseases were rampant. The importance of clean water was indeed known. For example, water was purified by boiling, but in practice, it was impossible to organize sufficient water purification for large populations. The Romans did take care of clean water for their armies, as the connection between clean water and the well-being of the troops was understood. But in fact, water was not drunk as much as it is today; instead, beer and wine, which were free from bacteria, were consumed. Wine was just usually diluted with water. Although the Romans had a magnificent aqueduct system that brought clean water into the city, due to the lack of sewage treatment, the prevalence of infections, and epidemics, the life expectancy was only 20–30 years. A significant change did not occur until the mid-19th century in Europe with the advent of the germ theory of disease, when the importance of hygiene and handwashing began to be understood. Actually, we have now gone to the opposite extreme from those times, as disinfection also kills beneficial microbes.

CLEAN WATER

In fact, obtaining clean water in nature, in the mountains, is not so easy anymore. In August 2022, the news reported that Earth's rainwater is no longer potable. Research carried out by Stockholm University and the Swiss Federal Institute of Technology had found that rainwater is no longer clean enough to drink as it is, whether the water is collected from the ice caps of Antarctica or the mountains of Tibet. These areas are considered pristine, yet they have PFAS levels 14 times higher than Western drinking water standards. According to the research, rainwater in no part of the world meets the safety limits for PFAS compounds.

PFAS compounds, or perfluoroalkyl chemicals, are hydrocarbon chains where hydrogen is replaced by fluorine. Due to the extremely durable carbon-fluorine bond, these compounds and the products made from them decompose very slowly in nature. PFAS compounds are found in many everyday products, including plastic packaging, frying pans, and non-flammable fabrics, and they enter human bodies most commonly through food. The presence of PFAS chemicals in rainwater has now reached such levels that collecting rainwater for drinking poses a health risk, leading to various health challenges. According to Professor Ian Cousins, the lead researcher, one should not worry about daily exposure from mountains, flowing water, or food: *"You can't escape it and should just get used to living with it. However, it's not a good situation if the exposure we get from the environment is not at a safe level."*

The shortage of clean drinking water is a major global problem. As of writing this book, the UN estimates that Earth will run out of clean water by 2050.

This is very sad. Waters not only contain the chemicals I mentioned but also constantly accumulate other chemicals, plastics, pharmaceutical residues, antibiotics, and pesticide residues. Annually, shocking amounts of drugs even in Finland, where wastewater is treated in purification plants, make their way into water bodies. An article in the Finnish pharmacy magazine *Farmasia* in Fall 2019 estimated that ibuprofen alone accounted for about 130 kilograms (286.6 pounds) annually entering Finnish water bodies, not to mention all other pharmaceutical residues! These residues enter fish and the soil, plants, animals, and also humans through water.

In Finland, waters are purified, but water purification is not at the same level everywhere in the world, and not everything can be purified. There are regular reports on the growth of ocean garbage patches. Fortunately, there are innovations and projects for water purification. For example, the project founded by Dutch 18-year-old inventor Boyan Slat, The Ocean Cleanup, ambitiously aims to clean the world's oceans of plastic debris. He got the idea for the project while diving in Greece, shocked by the amount of plastic in the sea. The project involves installing floating boom systems in the sea to direct ocean plastics to them, and then transporting them to different countries for recycling. Recycled plastic is used as raw material in industry. You can buy various products, such as a cell phone case or a swimsuit, made from ocean plastic debris.

Tap water is also treated with chemicals, such as chlorine. For example, in the USA, fluoride is added in many places to prevent the growth of harmful bacteria in the mouth and to prevent tooth decay. However, fluoride and its excessive intake have many disadvantages. Excessive fluoride has been linked to an increase in ADHD, neurological problems, and concentration issues, in addition to fluorosis. Fluorosis appears as erosion of tooth enamel and darkening of the teeth, and in severe cases, changes in the skeleton and calcification of ligaments.

In Helsinki, Finland, the water of the capital region comes through the Päijänne tunnel from Päijänne lake, and chlorine (according to Helsinki Region Environmental Services HSY, 0.35–0.50 mg/l) is pumped into the water network as bound chlorine, i.e., chloramine. The purpose of chlorine is to prevent microbial growth in tap water as it travels through the pipes. The amount of chlorine decreases as it oxidizes in the water network, and in the samples taken, it has been about 0.03–0.5 mg/l.

We are beginning to understand that chemical and pharmaceutical residues in water are an even more significant issue than previously thought. Earlier in this book, I discussed how water seems to have some kind of memory: It can begin to mimic compounds it has come into contact with, producing the same effects as these original compounds. This could mean that water, with antibiotic and chemical residues, might produce the same effects!

How then to prevent harmful substances from entering waters? The problem has been known for years, and attempts have been made to solve it, but so far, no viable solutions have been presented. A major change is needed, which also requires each of us. Firstly, the unnecessary use of medications and harmful chemicals should be reduced. The less these are used, the less, of course, they end up in wastewater. If there are preferably natural and otherwise side-effect-free alternatives to chemical medications, they should be used primarily. Naturally, there should be a focus on eliminating the cause of disease through lifestyle changes rather than medicating symptoms. This also means a shift towards integrative medicine, where Western medicine combines with complementary treatments. It is high time to treat humans as a whole, consisting of body, mind, and soul – both material and immaterial.

New solutions for water purification also need to be considered. The Exclusion Zone (EZ) phenomenon could also be utilized more extensively in water purification — the EZ area repels impurities, minerals, and even salts. Water purification devices based on this discovery have already been designed.

Marcel Vogel and water purification with the help of quartz crystals

Marcel Vogel was a researcher and inventor at IBM for nearly 30 years, specializing in phosphor technology, liquid crystal systems, luminescence, and magnetism. Among other things, Vogel studied the therapeutic application of quartz crystals and in this context, also researched their effects on water. He discovered that he could alter the structure of water using crystals that were cut, treated, and charged in a certain way, thus changing the water's properties. In his experiments, Vogel found that such treatment reduced the surface tension of water, increased its electrical conductivity, and changed its pH by 3 units, thereby increasing its acidity or alkalinity. Infrared (IR) and ultraviolet (UV) spectra also showed two new peaks, indicating structural changes, including the stretching of molecular bonds, meaning that the water gained additional energy and information. The freezing point of water appeared to drop by up to 30 degrees Celsius. A significant number of molecules also seemed to organize into new microclusters or molecular chains, systematically repeating the same pattern. Thus, the water appeared to become a liquid crystal system capable of storing information. Vogel's intention was to utilize these properties for information storage, as well as for water purification, among other things.

Could this create a connection to how stones affect us? For example, it has been shown through Kirlian photography that a person's energy field intensifies when they are in contact with crystals, suggesting that crystals enhance the body's energy. The energy of crystals combines with the body's energy, thus strengthening it.

HOW TO MAKE YOUR OWN ENERGIZED WATER

How, then, can the water used for drinking and watering also become structured, energized, "living" water?

Although EZ water has the same composition as tap water, it possesses different physical and chemical properties. EZ water can mainly be identified only with laboratory equipment, but the conditions under which such water can be produced are known. Some say that EZ water tastes softer, like spring water, and quenches thirst more quickly.

In a healthy human, the internal electrical potential of cells is negative. Cells contain free electrons. There is more EZ water inside the cells than outside, and cells both need and produce EZ water. EZ water hydrates better with the same amount of liquid than regular water and maintains the cell's internal mineral balance better. By drinking EZ water, thus, you're energized physically as well.

Here, I'll summarize the ways one can generate EZ water:

- **Using light, or electromagnetic radiation, whether it be visible light, UV light, or IR radiation.** IR radiation is the most effective of these, creating EZ water whenever there is a hydrophilic surface near water. Sunlight, especially containing IR radiation around 3100 nm wavelength, is good for this. It's beneficial to spend time in sunlight, in moderation of course, or to let drinking water energize in sunlight. IR saunas and regular saunas, where thermal radiation heats the skin's surface layers, produce more EZ water. However, instead of just influencing the surface layer, it's better to drink EZ water. Dr. G. Pollack studied water samples with a UV spectrometer and noticed that in the UV radiation area of 270 nm, EZ water absorbs, or receives, light. The more the samples absorbed UV light, the more EZ water the sample contained. Water samples from the Ganges river and Lourdes in France have also been studied, showing peaks in the UV spectrum precisely in the 270 nm area, indicating these sacred waters seem to contain a lot of EZ water.
- **Finding it at the interface of ice and liquid.** When water freezes into ice or when ice melts into water, the change occurs through EZ water (ice –> EZ water –> regular water and vice versa). Thus, EZ water is actually an intermediate state. The meltwaters of glaciers, for example, mountain streams in spring, are one of the best sources of such water. According to a study by E.V. Giudice et al., up to 85% of water droplets forming aerosols in the waterfalls of the Austrian Alps were negatively charged. Perhaps the

positive effect of ice swimming on health is partly due to the higher EZ content in water near the freezing point? And of course, it's due to grounding too, since the body "recharges" and fills the possible electron deficiencies from the water. In my own studies, the different pH of water melted from ice was also observed. Perhaps this too was evidence of EZ water?

- **Deep springs**. The deeper the water comes from, the better, as pressure also aids in the formation of EZ water.
- **"Vortexing," "pranaizing" or putting water in a spinning motion or aerating it by introducing air through shaking or pouring from one glass to another.** Sun-vortexed mountain stream water is surely one of the best sources of EZ water. This is said to return the water to the "memory" of its movement in nature, instead of traveling through pipes. Vortexing can also be done with a magnetic stirrer, but the effectiveness of this is somewhat questionable, as water is a diamagnetic substance and therefore cannot retain magnetism. Diamagnetic substances also repel magnetic fields. The purpose of magnets is just to set the water in motion. Pollack says that vortexing restores the water's memory of its natural behavior after being trapped in pipelines.
- **With stones.** Stones and crystals have been used in healing for hundreds of years. While visiting Europe's only currently operating gem mine in Luosto, Finland, I learned about Hildegard of Bingen's amethyst water. Hildegard of Bingen was an abbess, writer, and mystic who lived in Germany in the 12th century and wrote several works from visions and theology to natural sciences. In her book *Physica*, she addressed questions of natural philosophy. In her three visionary books, she described her visions, which she experienced in her waking consciousness. Visions were so vivid, *"as if seeing them with her five senses in the light of God."* Hildegard was known for her healing skills, using various herbs, tinctures, stones, and water for healing. For example, she is said to have healed a blind person with Rhine water. She advised using amethyst water for skincare and keeping the skin youthful, removing "spots from the skin," likely referring to age-related pigmentation spots. The stone had to be moistened with one's saliva and then rubbed on the affected area. Amethyst water removed itching and swelling caused by insect stings. Amethyst was also used for

protection against hostile people and to repel unwanted guests: snakes, spiders, and lice stayed away.

- **Adding minerals.** Water can be enriched with minerals such as Mg, Ca, Zn, which increase the EZ zone. Probiotics also produce EZ water. In a study by A. Sharma et al., the EZ zone increased steadily to a concentration of 0.075 g/100 ml, after which the EZ area began to decrease.
- **Using water with a pH near 7.** The water should not be too alkaline (pH above 9) or too acidic (below 4), as this reduces the amount of EZ water.
- **Using glass bottles rather than plastic ones.** H_3O_2 water (EZ) requires a hydrophilic surface to form, and it's best if the water is in direct sunlight. Glass is hydrophilic, but plastic is not. In the Ho'oponopono technique from Hawaii, water is placed in a blue glass bottle in the sunlight for at least an hour. Water is usually colorless, but among colors, it is closest to blue. The purpose of the blue color is to strengthen the water, incorporating the memory of its purest essence and origin, purifying the water of all negativity. Thus, this process shows respect to water by keeping it in a blue container while also charging it. The Ho'oponopono technique includes speaking the phrases: *"I'm sorry. Please forgive me. Thank you. I love you."* These sentences carry a strong intention and can be used to face any challenge. Joe Vitale and Dr. Ihaleakala Hew Len discuss this technique and their experiences, including the healing of patients suffering from mental health issues, in their book Zero Limits: The Secret Hawaiian System for Wealth, Health, Peace, and More. Ho'oponopono includes the same idea as Reiki: *By healing your own thoughts and beliefs, you heal everything around you.*
- **With sound.** Ultrasound induces the growth of EZ water with a delay. Ultrasound can also be used for water purification. Inez Hua (Purdue University) found in her studies that the frequency of 358 kHz was very effective – an alternative worth considering instead of chemical purification.

There are also devices by various manufacturers available for purchase that can produce structured, energized water. Unfortunately, it is often hard to find detailed information or research results about these devices. However, this does not mean that the devices do not work or that they do not produce the best EZ water for cells.

All water is actually in some way energized and organized, and most important for us is simply to drink enough water. It's also good to remember that the body naturally turns ordinary H_2O water into EZ water (H_3O_2), thus producing the EZ water it needs. According to Pollack, however, the body more readily accepts energized water than ordinary water. In a pilot study conducted by Korotkov, 15 individuals drank a liter of energized water daily for a month. In the experiment, a commercial product that transforms water into EZ water through a special glass surface and light was used to energize the water. The experiment found a clear difference compared to the control group: In the group that drank energized water, participants' body fat percentage decreased more, as did the body's (endogenous) state of toxicity. Water-salt metabolism, kidney function, and the body's adaptation and stress response, on the other hand, improved.

ENERGIZED WATER FROM PLANTS

Whenever possible, it is of course advisable to utilize the natural EZ water produced by plants: fresh and healthy plants contain a lot of energized water. The best sources are all organically grown and cultivated green plants, which are eaten soon after picking. Thus, the very best option would be to grow the plants oneself.

Nature has many plants that are particularly rich in water, such as cucumbers, watermelons, and aloe vera. A green smoothie made from fresh vegetables appears to be an especially good source of EZ water. Pollack has also studied chia seeds, which seem to generate a significant amount of EZ water in the gel layer that forms around each seed when they are placed in ordinary water. Ghee, a clarified butter used in Indian food, also transforms ordinary water into EZ water. The research group of A. Sharman mentions, among others, tulsi (a type of holy basil belonging to the Indian basil family), turmeric, and coconut water, all of which increased the EZ zone in their study. They also examined the effect of the glyphosate-containing herbicide Roundup on the EZ zone: the EZ zone decreased immediately with an increase in concentration.

Various energizing devices have been used to produce irrigation water for plants and drinking water for animals. Plants grown with energized water appear to grow faster, healthier, and with less need for fertilization than those watered with ordinary water. Such studies are of course interesting, as animals and plants cannot be thought to "believe" in the method's effectiveness in the same way humans can. Sometimes it is suggested that the user must believe in the method for it to work.

The concept of *"biohacking"* is relevant here. Biohacking aims to create the best version of oneself, helping the brain and body to function optimally. Often, biohacking is about the "maintenance of the machine," that is, maintaining one's physical existence. Depending on the person, this could mean wanting to lose weight, improve brain function, or have more energy. Biohacking could also be seen as drinking the most optimal EZ-rich water and eating a special type of food to feel as good as possible. The next level of biohacking, after focusing on the physical body, is paying attention to one's mental aspect – how to live in this body, the home of this life, in the best way for the soul.

HYDROGEN WATER

Drinking hydrogen water can also be considered as biohacking. Hydrogen water, also known as hydrogen-rich or hydrogenated water, is regular water with molecular hydrogen gas (H_2) added to it. According to some recent studies, hydrogen water appears to have antioxidative and anti-inflammatory effects, meaning it can neutralize harmful free radicals in the human body, indicating potential benefits for various diseases and conditions.

There are different ways to add hydrogen to water. Water can be hydrogenated by dissolving molecular hydrogen gas into water under elevated pressure. Hydrogen water can also be made using electrolysis, in which the cathode and anode electrodes are separated by a semi-permeable membrane. The water at the cathode produces electrolyzed reduced water (ERW), also known as alkaline ionized water, while the water at the anode produces an acidic water. The membrane prevents the two waters from mixing, allowing each to maintain their respective pH. ERW contains hydrogen gas.

Preliminary studies have been encouraging — hydrogen-rich water seems to contribute to various aspects of health, including physical endurance, liver function, cardiovascular disease, mental health, COVID-19, oxidative stress, allergies and anti-aging research. However, as with many early and forerunning studies, well-designed studies in humans with larger sample sizes and long-term trials are needed, to confirm the benefits.

MY WATER ROUTINES

I've already mentioned that I prefer to drink water from a mountain stream or spring water whenever that is possible. However, in the city, one often has to settle for filtered tap water. Filtering removes particularly chlorine and – in many parts of the world – fluoride residues, which prevent the necessary charge zone (needed for EZ water) from forming in the water. Neither is good to consume even in small amounts. For example, chlorine also kills beneficial microbes in the body. Sometimes you can really taste the chlorine residue in the water. Fluoride, on the other hand, can cause fluorosis, enamel damage to teeth, and in severe cases, damage to the skeleton when used excessively. Water filtration also removes copper that may detach from the pipes. Generally, the water filtration devices sold are based on activated carbon filtration, meaning the water filters through small micro-charcoal beads, with the charcoal capturing harmful substances.

When I pour filtered water into a glass, I try to *"pranaize"* or vortex it. Pranaization can also be done by pouring water from one glass to another a few times. At the same time, I seek connection with the water and send a loving intention with the thought to activate and restore the water's power, if, for example, it has diminished while passing through pipes. I treat the water for a moment depending on the situation. Treating means especially seeking connection with a background of grateful and loving feeling – a feeling of how we are one. I consider treating water especially important when someone is sick. All of this, of course, also applies to food.

Have you noticed how animals prefer to drink moving water? Perhaps this is fundamentally about the same thing; vortexed, oxygenated water is better. It would be challenging to conduct research on this. At least I don't always understand my cats' choices — I thought they preferred moving water until I saw them both drinking standing bucket water. Maybe this was the novelty attraction. The bucket was metal, not plastic like the moving water fountain. Maybe it forms EZ water better? For water storage, a glass jug is the best, as the plastic surface is worse for EZ water due to its water-repelling properties.

If possible, I take the water jug into the sunlight. The best is when there are as few plastic or glass layers as possible between the sun and water, ideally none, because the sunlight should fully penetrate the water, and for example, window glass filters this.

Then, does the temperature of the water matter – is it better to drink hot or cold water? In the mornings, I've taught myself to drink hot, nearly boiling water, to which I add just a little lemon juice. I drink about 3–4 deciliters, 10 to 14 fluid ounces, of this water. Yogi Ramacharaka writes about the benefits of drinking hot water in his book *Water Cure*: *"It acts along the lines of washing out the stomach; relaxing congested conditions there; loosening up the mucous accumulations, and dissolving them that they may be carried out of the system; and finally as a mild invigorator and stimulant."* Hot water thus wakes the body up to function well for the day. On hot summer days, however, I do not necessarily drink hot water. The most important thing is, of course, to listen to what feels best for oneself and to drink enough water throughout the day to hydrate the body well. Ice cubes are indeed a great idea for summer days and, of course, other times too, since there is EZ water at the interface of water and ice.

Sometimes I drink gemstone water, which is water in which I have added a dozen small crystals. My late Reiki teacher Sari McGlinn always offered gemstone water in her courses and shop, and my favorite was sodalite-quartz water. I've also drunk this water while writing this book – Sari mentioned it as clarifying and enhancing intuition. I find it gives me a more confident and relaxed feeling. Sodalite is a blue and white stone believed to be related to emotions, intuition, and deep understanding, and to seeing things more clearly and broadly. Quartz is transparent and generally associated with spirituality and spiritual growth, seeing things more clearly. However, please note that some sources indicate that sodalite is toxic and should not be placed in drinking water. There are also many other crystals, such as malachite, fluorite, pyrite, stibnite, actinolite, azurite etc. that should not be mixed with water due to their toxicity.

TREATING WATER

For thousands of years, different cultures and religions have performed various ceremonies to give thanks, strengthen their connection with water, and seek balance and harmony with water, ourselves, and the entire cosmos. All of these ceremonies could be said to be about treating water. We all hope to drink the best and most vital water — the very best for us. We should constantly treat water — send it good thoughts, as thoughts and feelings have an impact, even on the material world.

How, then, to treat water? What do I do?

Water can be treated using either a short or long method. Anyone can treat it, even without any knowledge of energy healing. The most important issue is the intention and seeking a connection with the water. I often place my palms around the glass or jug of water, although that's not necessary. Holding hands around the object to be treated helps form a connection. I breathe in deeply a few times through my whole body and, with each breath, seek a connection within myself and with everything — as if strengthening the connection to an all-encompassing and permeating network. I imagine in my mind the memory, the feeling of clear, flowing, refreshing, sparkling water, and imagine the water I am treating in the same way. I bring, in a way, the future or a memory from the past into this moment. There is no time, no distance, everything is in this moment. I charge the water with gratitude, love, joy, and happiness. I treat it as long as it feels right.

Here's a variation of a water ceremony for treating water using a longer method. You can perform the following ceremony right away; all you need is a glass or bottle of water.

Ceremony for Vital Water

If possible, do this outside in nature, preferably near natural water:

• Now go get some water in a glass, unless you already have a glass or bottle of water in front of you. Preferably glass.

• Observe the color or colorlessness of the water, feel the coldness of the glass or bottle's surface. Smell the water as if you were drinking wine. Taste the water, hold it in your mouth for a moment feeling its texture, coldness, and tasting its flavor.

• Then place the glass in front of you, on the table or ground. Close your eyes and seek connection with the water. You can hold your hands on either side of the glass, but do not touch the surface of the glass or bottle now. In seeking connection, initially use the senses you just experienced, what the water looked, felt, tasted like, and so on. Feel how this water is also inside you, as you just took a sip from the glass.

• The water you drank interacts with your own tissues, cells, and the water in them. They exchange information. In this interaction there are also the journey of the water you drank, how it ended up in your glass, where it has been, what it has "seen," "experienced," and all that you have encountered in your lifetime. Through water, we connect to a vast network spanning the universe, where water circulates in different forms, from one task to another. It has always been and will always be, long after us. Read these words in your mind or out loud to the water:

How it was once born, along with everything else at the beginning of time.
How it trickled down from the clouds as water, snow, even sleet.
How it once merrily flowed in mountain streams.
How it froze into massive turquoise ice masses at the south and north poles.
How it traveled in deep blue, lightless caverns at the bottom of the sea – places no human has ever been.
How it "saw" into our innermost being and knows our deepest essence.
How it contains all knowledge, all information, the same that is in you and me.
The same all-encompassing, permeating, eternal, and pure love.

• You can recall your experiences with water. In your mind, you can travel along your timeline, your spiral of time, to the past, bringing up those moments that have been meaningful. Maybe you notice how water appears differently in these. It is mysterious, a purifier, healer, steam, liquid, ice, hydrator, thirst quencher, ancient, always moving, joyful, living, life-bringing, growth-bringing, manifesting, constructing, sometimes a cruel destroyer, powerful, loud, artist, magician, changer, changing, teacher, gift, and surely much more.

• Keeping your palms still on either side of the water, you can send an intention, charging the water, so you get from it all that you need right now, in the best possible way: Gratefully, lovingly, selflessly, listening, hearing, and receiving.

• Releasing all obstacles, surrendering to feelings, even allowing them to flood. Take time, accepting, allowing, breathing. Speak to the water – to the messenger of life, love, and healing – with your whole heart, mind, soul, and body.

• As you treat the water, allow it to have permission to treat you. On the other hand, treatment is not even necessary, because everything is already here, since time does not exist; it is merely an experience. Gradually open your eyes and return to this moment and place. Finally, you can also recite Masaru Emoto's water prayer:

WATER WATER WATER WATER
We are sorry.
Please forgive us.
We thank you.
We love you.

Sip the water as you continue reading the book and enjoy its vitality, healing, and refreshing effect.

CHAPTER 12. THE METAPHYSICS OF WATER

The nearby mountains bring us
precious water from their springs,
giving refreshment, success, well-being,
and great courage.
– Sign at the fountain in the village of Boppard, Germany.

~

I'm sitting on a river cruise on the Rhine River – we're here on a summer vacation trip. The boat is moving downstream toward the Lorelei. Lorelei is a 433 ft (132 m) high steep cliff on the bank of the Rhine River near the village of Sankt Goarshausen in central Germany. At the point of the cliff, the river is at its narrowest, being only 370 ft (113 m) wide. Due to the strong current and underwater rocks, there have been several boat accidents over the years. In the past, whispers could be heard around the cliff, apparently from the combined effect of a waterfall and the rocks. It was spoken of as the whispering rock, and there are legends about how the beautiful golden-haired maiden Lorelei, having been disappointed in love, now called sailors to their death in the same manner as the Sirens of Greek mythology.

The river boat is just passing a spa – here the water is particularly mineral-rich, and the spas are popular for treatment of rheumatism, muscle pain, skin problems, and other ailments, or just for relaxation. In addition to the healing properties of water, the spas also utilize ancient knowledge of energies and the healing properties of stones in their treatments.

On our trip here, I have an empowering dream. I am on soft white sand by the shore of the sea, at the foot of a palm tree. The sea is turquoise blue, and toward the horizon, its color darkens and deepens. I see a dolphin – it comes from the sea to me and invites me to swim with it. Further out at sea, there's a rainbow, and we swim toward it with the other dolphins. In the dream, I think all this is absolutely impossible, yet I don't realize it's a dream. All the colors of the rainbow shine brightly as we arrive at it with the dolphin. I can choose a color, and we travel up the rainbow in the chosen red color first, then down. As we descend, huge waves come from the sea. At the same time, it feels as if the dolphin and I are one, and then the dolphin has disappeared – or not – I realize I am in the form of a dolphin, swimming in the sea free and happy. I wake up.

~

Dreams are messages from the unconscious to the conscious mind. I have kept a dream journal for a long time, learning to interpret the symbols in dreams. They can sometimes be peculiar, and often only the dreamer can best interpret their dreams. One must learn the language through which the unconscious mind communicates. Sometimes the symbols are more collective, and sometimes they are such that only the individual can understand – for instance, the time I dreamt of a blue angel sculpture that fell from the wall to the floor, and a piece broke off from one of its corners. Later, I figured out what my dream was related to. At that time, we had a blue car, which we called Gabriel because of the letters in its registration number. The day after the dream, I reversed the car in such a way that its rear corner was damaged. The blue angel was our car! If only I had realized this earlier and been more careful!

I search for the symbolic meaning of my dolphin dream. The sea is related to freedom. The dolphin, on the other hand, symbolizes joy and brings it to its surroundings. Even in captivity, dolphins are playful; they are also wise. They are associated with life energy, power, and lightness, playfulness. The dolphin teaches us to let go and to plunge in, not everything needs to, nor can be controlled. They teach us to breathe and to find connection. They also seem to be messengers, bringing a message of an important mission that unites many.

When we returned from the trip back to Finland, Finland's borders closed, and the world shut down again due to COVID-19. We are together in captivity, and the dream takes on a completely different meaning than I initially imagined.

~

Metaphysics seeks to understand the world from a level even deeper than physical reality. It contemplates the purpose of everything, the essence of being. The metaphysics of water, therefore, means a deeper understanding of water—a experiential, collective, ancient knowledge. We are actually moving to a level where we also experience how water feels to us.

Sometimes when I'm in water, I feel that water knows everything. It's hard to explain, but in a way, there's a feeling that I can never grasp or understand it with the conscious mind. The connection occurs on another level. Perhaps the water in me is talking to the sea water? Perhaps there is an exchange of information happening within us in different ways?

Water is difficult to control – it symbolizes the ultimate uncontrollability of reality. Water often symbolizes flow, the progression of things, life, and life energy. It is related to purification, cleansing, purity, and even holiness. Emotions are ever-changing, and so is water. It is also mental power, potential, constant changes, renewal, and variability, sometimes also weakness, and a reminder that you are capable of much more. Water and the sea have symbolized women, motherhood, and femininity across different cultures through the ages.

Metaphysically speaking, water also represents the collective mind – collective consciousness and its changes. Reality is constantly changing and will not return to what it was before.

INTERPRETATIONS OF WATER IN DREAMS

In Sigmund Freud's *The Interpretation of Dreams* (*Die Traumdeutung*, 1900), water symbolizes the oceanic experience of broader unity, although sometimes water can also represent a frightening state of regression. For C.G. Jung, water is one of archetypes of the collective unconscious. It is the most common symbol for the unconscious. In The Red Book, Jung states: *"Dreams are the guiding words of the soul. Why should I henceforth not love my dreams and not make their riddling images into objects of my daily consideration?"* Dreams can be a good aid in gaining new perspectives on life and upcoming events. According to Jung, some dream symbols are collective, symbolizing things that are similar for everyone. However, some symbols are more individual, and the language with one's inner self is formed as a collection of these. I have compiled a dream-symbol book for myself, and in it, water symbolizes the following things:

- **Running water, such as a stream, river, or larger current**. A lot of energy. It also represents time and the progression of events. A small stream means joy and lightness but also slow or calm progression of life, as only a little water flows forward. The larger the stream, the more forcefully and quickly things progress and happen, and with more energy.
- **Calm water**. Calmness is a state of mind. Things are fine right now, and perhaps they even need to change, because in balance there's no progress, and sometimes one must go to extremes for growth to occur.
- **Storm**. Rapid change of matters. Fear can guide, be wary of that. Expectations can be different from reality; you may fear in vain.
- **Clean water**. A positive sign related to different issues – can symbolize well-being, pleasure, happiness, success, clarity. Dirty water, on the other hand, symbolizes more negative things and warns of danger, obstacles, challenges.
- **Drinking cold water** is a promising sign – means success and achieving goals. Hot water is not so promising a sign, pointing to something negative or at least the need to be on guard.
- **Swimming in cold or cool water** is similarly a good sign and can also relate to new beginnings. Correspondingly, hot water warns. The hotter it is, the more it warns. Boiling water means something is coming to the surface, issues are emerging.
- **Flood**. If water floods into your home, something threatens your well-being or health.
- **Walking on water**. Overcoming the material, the spirit's victory over matter. Capable of achievements that no one could believe. You can achieve wonders. A good sign: joy, happiness, success, you are helped and supported.
- **Rain**. For some, rain is a bad omen, but for me, it means purification and getting rid of worries. If rain is associated with fog, something is unclear and may predict challenges. A downpour can mean either rapid progression of matters or that something negative is coming your way, especially if you stand in the rain without an umbrella or other gear. If you

also see these, it means you are prepared for upcoming challenges or that you should prepare better or maybe also ask for help. Rain also symbolizes growth, fertility.

• **Sea.** A sign of freedom. Waves bring new winds. The more waves and foam, the more emotional turmoil and events.

Jung and the client who consistently dreamed of water

In the book *The Archetypes and the Collective Unconscious*, Jung recounts a letter he received from someone who described having recurring dreams about water: *"Almost every time I dream it is about water: either I am having a bath, or the water-closet is overflowing, or a pipe is bursting, or my home has drifted down to the water's edge, or I see an acquaintance about to sink into water, or I am trying to get out of water, or I am having a bath and the tub is about to overflow,' etc."*

Jung's interpretation of the dream: *"Water is the commonest symbol for the unconscious. The lake in the valley is the unconscious, which lies, as it were, underneath consciousness, so that it is often referred to as the 'subconscious,' usually with the pejorative connotation of an inferior consciousness. Water is the 'valley spirit,' the water dragon of Tao, whose nature resembles water – a yang embraced in the yin. Psychologically, therefore, water means spirit that has become unconscious."*

Perhaps, the unconscious is trying to communicate something important. Maybe it knows something that the conscious mind does not? It could be, for example, a warning or a directive to do something. The final part requires personal interpretation, regarding what this could mean in the case of the dreamer.

WATER AND THE OTHER ELEMENTS

In the founding philosophy of many cultures, including Ancient Greece and China, the world is seen to be made of four or, more accurately, five elements, one of which is water. It must be stated from the outset, however, that this concept is entirely different from the immutable elements of Western thought, even though the term "elements" is sometimes used in both contexts. A more accurate description might be "influencers" depicting the cyclical variations of nature. Indeed, the conceptual frameworks described in this chapter are often easily dismissed with the assumption that we now have better ways to describe the world. However, these frameworks contain thousands of years of observing the world and its phenomena.

Plato describes in his *Timaeus* how the world consists of four elements: earth, air, water, and fire. These are formless, mixed into each other, and in constant motion. To describe the elements, Plato used solids: earth was represented by a cube, air by an octahedron (8-faced), water by an icosahedron (20-faced), and fire by a tetrahedron (4-faced) (Figure 21). Each of these polygons consists of triangles and can thus be divided into the triangles that formed them.

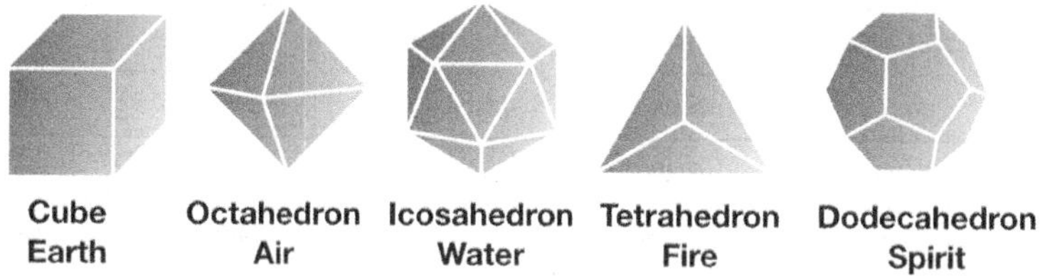

Figure 21. Platonic solids.

Thus, the elements could be transformed into each other, since triangles could be used to compose any element. This idea was later foundational to alchemy and, subsequently, chemistry. The contemporary materialist worldview incorporates a similar notion of reductionism: the material world can be reduced, or decomposed, into ever smaller parts.

Plato also proposed a fifth element, or quintessence. Represented by the dodecahedron (12-faced), quintessence was described as an essence and space was said to consist of it. It has been thought of as ether, and nowadays even spirit or consciousness. In Hindu philosophy too, the world was composed of these five elements, with ether being the fifth. Water, among these five, was the one with a variable, changing state.

In Chinese philosophy, matter is a combination of two states of *qi*. *Qi*, the life energy, is akin to what is referred to as prana in Sanskrit and mana in Hawaiian. *Qi* differentiates the living from the non-living.

Chinese philosopher Xunzi (3rd century BCE) described it as follows: *"Water and fire have qi (vital essence) but no life; grass and trees have life but no consciousness; animals have consciousness but no righteousness; human beings have qi, life, consciousness and righteousness; therefore, humans are the most precious (thing) in the world."*

When the states of yin and yang in *Qi* combine, matter and its five elements are created: water, fire, wood, metal, and earth. These are in a constant state of change and transform into each other: wood –> fire –> earth –> metal –> water –> wood –> and so on. Thus, each of the elements contains a certain property of *Qi*. Different characteristics associated with the elements are listed in Table 4.

	Water	Air	Fire	Earth	Spirit
Properties	Cold, moist	Warm,moist	Warm, dry	Cold, dry	-
Sex	Feminine	Masculine	Masculine	Feminine	-
Activity	Passive	Active	Active	Passive	-
Color	Blue	Yellow	Red	Green	Violet, white
Planets	Moon, Venus	Jupiter	Sun, Mars	Saturn	-
Horoscope signs	Cancer, scorpio, pisces	Gemini, libra, aquarius	Aries, leo, sagittarius	Taurus, virgo, capricorn	-

Table 4. Characteristics associated with different elements.

Chinese philosophy places emphasis on the idea of unity and the fundamental nature of consciousness. Matter arises from consciousness. This conclusion can also be reached through modern physics. In *Hyperreality*, I wrote: *"Matter, experience and consciousness would seem to have a connection. We shouldn't consider them as separate, but instead by studying matter we learn more about consciousness, ourselves and reality."*

Hildegard and the Interpretation of Elements

During the Middle Ages, many healers used knowledge of the elements to help those in their care improve their health. Hildegard of Bingen, whom I previously mentioned regarding amethyst water, also utilized the teachings of the elements in her work through the so-called humoral theory, or the four humors. This theory described a person's character and physiology through four bodily fluids: blood, bile, phlegm, and black bile. The element of earth had the same properties as black bile. It was cold and dry. If black bile dominated a person's body, they could suffer from melancholy, which was also another name for black bile. Many artists were prone to melancholy; it was both a state of mind and feeling, and in its extreme, also a disease.

When the elements were in balance, a person was healthy; imbalance led to diseases.Hildegard states in her work (freely translated) that as long as the flow of humor in a person functions properly and maintains warmth, moisture, blood, and flesh, then the person enjoys good health. But as soon as the flow becomes too much and suddenly, it causes illness and can lead to death.

According to Hildegard, fire strengthens, earth is particularly related to life energy, air is flexible, and water moistens and nourishes. None of these can manifest without the others; all flow together. Similarly, all of creation is one and interconnected, linked to the elements. Everything has its purpose and task.

ABOUT QI AND WATER

Qi – the life energy – seems to have a connection to water. In many complementary and Eastern healing practices, there's an emphasis on understanding water in its various forms. The *"living water"* mentioned in the Bible could also be a description of *Qi*: *"Whoever believes in me, as Scripture has said, rivers of living water will flow from within them"* (John 7:38). This is said to refer to the Spirit that those who believe in him would receive, essentially *Qi* and its strengthening. It's writing about the same thing, just in different terms.

In the Finnish book *Veteen kirjoitettu*, in English *Written in Water*, it is said: *"Cranial osteopaths believe that water bound to the body's collagen tissues mechanically, biochemically, and energetically connects different parts of the body. Osteopathic treatment is believed to transform this water into a semi-solid gel, upon which healing procedures can be applied. Osteopaths call the effective forces of the procedure the 'breath of life.'"* Water, therefore, is a unifying factor, and it's spoken of in the same way as the life energy *Qi*. The book also states that all Eastern health care practices that use water and its unknown energy in some way seem to work. Such practices include Reiki, qigong, and taiji.

Perhaps the properties of water are indeed the properties of *Qi*? In water, *Qi* manifests in its purest form, for water is boundless. In us humans, it manifests in each of us in our own form, for are we not just mirrors, reflecting reality with our being?

Indeed, the connection between water and *Qi* has been explored. Ulisse di Corpo and Antonella Vannini (2010) have suggested that the formation of hydrogen bonds could create a mechanism by which consciousness affects matter. Through water, consciousness could link the particle level to the macroscopic level. This might be the reason why the water molecule is so essential for life. Perhaps because of its unusual properties, water has become the *"molecule of life."* Water may be sensitive to bioenergy (*Qi*) because it is sensitive to consciousness, which could also explain why water is an essential part of all biological life.

Water, air, fire, earth, and naturally spirit are essential elements for life. Understanding them has become obscured in current Western society. Yogi Ramacharaka states that water, air, and the sun are nature's three greatest gifts to mankind. They could almost entirely eliminate disease if humanity used them correctly. Perhaps the courage to explore unexpected directions could be the solution and provide answers?

CHAPTER 13. IN DEEP WATERS

"A great ship
asks deep water."
– Poet, priest and theologian George Herbert (1593–1633)

I'm once again swimming in cold water, it's November. The sky is dark, clearly indicating that a storm is coming. The overall appearance everywhere is gray, although the leaves still hold their color. The water temperature is below ten degrees Celsius, 50 °F. I swim as I do in the summer, 50 strokes toward the open sea, then I listen to my gut feeling and decide whether to continue forward or turn back. Water talks to me. Or actually, it doesn't speak but communicates without words. The water feels gentle and at the same time powerful, and I sense how powerless I am. There's also a feeling that everything is going as it should. In my mind, I see an image of myself sinking deep into the dark water and seeing light above me when everything else is dark blue, almost pitch-black. I dive.

I notice a deep silence, even though I only dive for a moment. My friend, the writer and diving instructor Elena Gushchina, who is also an artist and paints pictures of water, mentioned this silence when diving. Silence, completely being in this present moment, and connection to water. It's difficult to find a similar experience anywhere else but under the water.

Water has a lot in common with space. Both are deep, unknown, infinite, eternal, all-encompassing, permeating everything. For me, the image of diving in dark, night-blue, velvety water is a symbol of a spiritual journey – of the courage to go on an exploration into the unknown. Water connects us humans, animals, plants, the earth, the entire universe, as does space. In dream interpretations, diving into water can signify achieving a connection with nature, but it can also mean threat, regression, and the disintegration of the self. This is mentioned in the book Veteen Kirjoitettu, in English *Written in Water*. In psychoanalytic interpretation, water represents the world of unknown and repressed truths. Often, water is also a symbol of creativity, embodying flow and the birth of something new.

To get answers, sometimes it's necessary to dare to go deep and in unexpected directions. Where will the water lead next? *"I am ready to dive exactly there, where it takes me,"* I write down when I return from my swim.

YOU ARE THE WATER – THE SACRED CUP

The symbolism of water and ancient knowledge is encountered in surprising places. Nowadays, Tarot cards are mainly associated with fortune-telling, forgetting that the cards are connected to various esoteric traditions, astrology, alchemy, Kabbalah, Christian mysticism, and numerology, among others. The origins of Tarot cards are not fully known, but they have been used for centuries as both a parlor game and for reading. One of the oldest sets of Tarot cards is the hand-painted Visconti Sforza deck made for the Duke of Milan around 1440. Antoine Court de Gébelin studied the history of Tarot in the late 18th century and concluded that the Tarot cards and their symbolism originated from the esoteric traditions of ancient Egypt, including the Book of Thoth. Thoth was originally the moon god, later also the god of wisdom, knowledge, writing, and science. The Book of Thoth was historically kept hidden in a temple in Egypt in a golden box, the key to which was held by the *"Master of Mystery."* It is said that for the reader of the book, reality unfolds in a way that is only possible for immortal beings.

De Gébelin created his own Tarot deck, on which modern decks are based, with the cards based on the number seven: There were a total of 77 cards plus the Fool card, making 78. The major arcana includes three times seven cards plus the fool, and each of the four suits contains two times seven cards: numbered cards from ace to ten and four court cards.

In *Hyperreality* I wrote about Jung and synchronicities – meaningful coincidences. Jung's view was that the Tarot card we choose unconsciously answers some inner need: something needs to be revealed to us, to come to our knowing, to become visible in this reality. Through cards we actually connect to the universal all-encompassing unified field of information, or what Jung called the collective unconscious.

Cards can be thought of as mirrors in the same way that the surface of water reflects an image into our field of vision. When you look at the surface of the water or a mirror, do you see what you want to see or what you truly are? Similarly, when you draw a card, your perspective is reflected in what you see in the card and how you interpret its messages. Gradually, you learn to understand the symbolic language and how to interpret what you see. Therefore, interpretation does not necessarily require cards; you can continuously make the same interpretations in life from everything you see and encounter. However, Tarot cards offer an easy auxiliary language to learn to interpret the messages your inner self tells you. We are not separate from the surrounding world, and we must learn the language to interpret reality.

Therefore, I delve into Tarot cards to learn what they can tell me about the mystery of water. There are several cards in the Tarot related to water and containing water symbolism: all the Cups (or Chalices) suit cards, from the major arcana *The Moon, The Star, Temperance, Death and The High Priestess*. Also, among the Swords suit cards, the water theme is present depending on the deck at least in cards 2, 5, 6, and 8, as well as in the Pentacles card 2. The following table describes these cards and how these particular cards could be interpreted. The description is based on the widely used Rider-Waite-Smith (RWS) Tarot cards, which were first published in 1909.

Card	Description	Interpretation
Cups	Cups in different ways	Emotions, feelings, and the relationship with other people and the world. When drawing a card, listen especially to your feelings and reactions.
The Moon	Two dogs or a dog and a wolf howl at the full moon in the countryside, two towers, a lobster emerges from the sea.	The tame and wild side. The lobster represents history and traumas that are passed down through generations. The Moon is a symbol of emotions, intuition, and femininity. The sea represents the subconscious and depth.
The Star	A woman on a pond, naked, pours water, with seven small stars and one large one in the sky. One foot in the water, the other on land.	In balance and daring to be who she is. The card represents hope, creativity, the flow of things (water). The revelation of things.

Death	A skeleton rides a white horse, armored, holding a flag in its hand with a white rose on a black background. In front of the horse are a priest, a young man on his knees, and a child. In the background, there is a stream of water.	Death is a card of transformation, signifying new beginnings. The flow of water represents the inevitable progression of things. The people in front symbolize different perspectives.
Temperance	An angel balances two cups in their hands, from which water flows in an unusual way. There is a stream at their feet, one foot on the ground, the other in the water.	Grounded, yet connected to the subconscious and emotions. All the tools are already available, and the path is visible; make use of what you have. Harmony and understanding.
The High Priestess	The High Priestess sits between two pillars, a crescent moon at her feet. The card itself does not contain water, but symbolically relates to themes of water.	A unifier of two different sides. The crescent moon relates to intuition and emotions, the subconscious, and femininity.
Swords: 2	A figure sits with two swords, blindfolded, with a rocky wave and a crescent moon in the background.	The figure does not listen to their feelings, does not trust in intuition or emotions.
Swords: 5	A figure gathers broken swords, apparently having won. In the background, a couple of figures, one of them crying.	Victory, but at what cost. Thinking only of oneself, unemotionally.
Swords: 6	A man, woman, and child in a boat, setting off on a journey, carrying swords with them. One side of the boat is calm, the other wavy.	Leaving is not always easy, but sometimes it's the only option. A decision based on reason, yet feelings suggest otherwise.
Swords: 8	A figure with eyes and body bound on the shore, swords stuck into the beach sand.	Bound, incapable of doing anything, at least that's how it feels. Unable to see solutions, just waiting.
Pentacles: 2	Ships balance on the waves in the background of the card.	The figure balances, with an unreal-looking sea in the background. Perhaps not everything is as it seems.

Table 5. Water-related symbolism in the Rider-Waite-Smith Tarot cards

I conducted a three-card spread with the intention of better understanding water. What could it tell us, first about its history? And now, what does it want to bring to our attention right now? What does the future of water, especially in the next hundred years, look like?

The history card was *The Moon*. This card represents the unconscious, intuition, feelings, and mysteries. The tame and wild sides of water, as well as

Picture 22 a, b, c. From the Visconti-Sforza Tarot cards (from 1440), The Moon, King of Swords, and King of Cups.

its long history, are revealed in the card. Perhaps knowledge or even traumas that have been passed down from previous generations to us? The RWS deck card I used shows two towers at its edges—a symbol once used to denote the gates to sacred and mysterious, unknown worlds. However, the card is most prominently dominated by a large full moon, which speaks to leveraging the unconscious mind and instincts. I can well believe that in ancient times, there was a stronger listening to the subconscious mind, living also in harmony with the water and the moon. The history of water is shrouded in mystery, it has also been viewed as something mysterious in the past.

The "Now" card was the *King of Swords*. In the card, the king sits with a sword in his hand, looking stern and strong, on a throne adorned with butterfly decorations, a cloak on his shoulders. His figure fills the card. The card looks exactly how I might imagine water to look at this moment, considering the current state of the world's waters. A lot of the water on our planet is being controlled and ruled over by "stern men". It's time to use our knowledge and experience, to change the situation, actions are needed!

The future card was the *King of Cups*. Here, a serious-looking king sits on his throne floating on water, holding a cup and a scepter in his hand. The waves ripple, and judging by the king's expression, the situation is demanding. The king in the card seems trustworthy and capable of wise decisions, with reason and emotions appearing to be in balance. He governs and waits for us to act in favor of water – for our own good. This could mean that, in the future people who rule over water might see water more with emotion, in a larger context and not only as a commodity. There is hope that they will understand the deep importance of water to our planet. Water will be one of the things hopefully uniting the population.

The future card also relates to synchronicity. I wrote down these words from the lyrics of Ajeet's song "*Haseya*" before I started writing this Tarot section:

"Rise up, my sisters, rise up,
We are the water, the sacred cup
It's in our hands that all life grows
It's in your dance, it's in your hands
It's in your love we rise above."
– Ajeet, from *Haseya*

Thus, we are water and cups. It is in our hands to ensure that life continues into the future.

ON THE WAVES OF REIKI

If everything doesn't go as you'd hope, usually the first thing that surfaces is an emotion: for example, fear, anger, or sadness. As time passes, the emotion often changes. Emotions are constantly changing, just like water.

Emotions are a signal from within. They tell you that either things are going well or something urgently needs to change. Emotions are also connected to the unconscious mind. They reinforce memories – through them, a deep memory imprint of the experienced event is formed. The event remains in memory, creating new neural connections or strengthening old ones while others may weaken. The emergence of various fears is a good example of this: because of fear, encountering spiders, flying, or going into water can feel insurmountable.

I was in swimming school at the age of four. At that age, swimming school is still quite a playful thing, but for the first time ever, I was in a test: Everyone who could retrieve a hockey puck from the deepest end of the children's pool would get to go to the large pool. I was the youngest in the group, but definitely one of the most competitive. I didn't want to be worse or different from the older ones. Of course this then backfired. Others were already fetching their puck from the bottom of the pool, and I tried too, but in the rush of diving, I forgot to hold my breath as we had been taught. Water went into my nose and lungs. I almost drowned. The instructor lifted me from the water and I was so frightened that I ran crying to my mother who was waiting and watching me in the cafeteria of the swimming hall. I still clearly remember that event, although it's been nearly five decades.

My attendance at swimming school ended there. I no longer wanted to go. Fortunately for me, soon after that event, I also experienced good moments in water, such as those at the summer cottage, which had a suitably shallow sandy beach for wading. Amidst water play, exploring the shore waters and its insects, I also learned to swim unintentionally. Good moments in the water replaced the unpleasant ones, and what remained most strongly in my mind was the experience of success and trust in water. However, it could have gone differently.

In *Reiki*, there is a specific symbol for treating emotions – Seiheiki. It is like a wave of water when I draw it over and over again. For me, it is envisioned as aqua blue, translucent, gentle wave that sweeps over me again and again. As I mentally trace its pattern, I also hear the calming sound of waves in my ears. It's the same sound as from a gigantic seashell, which my grandfather put to my ear as a child, saying it contained the ocean's waves.

The image of healing waves brings me trust and hope, that's how I feel it in my mind. I also feel that my body and emotions are increasingly just pure energy, and it's also possible to influence that energy. The material is only temporary and changeable. The material world seems to melt into liquid, which is also easier to shape. I am in a Reiki sea, where I let myself float on the water's waves – I can let go.

Figure 23. Seiheiki waves.

Healing Emotions the Reiki Way

If you experience negative emotions, it's a signal from within that not everything is okay at the moment. Something needs to be done to prevent the situation from continuing as it is for any longer.

Understanding emotions is especially important because they guide our actions. However, it's often forgotten that thoughts and emotions are intertwined, and we usually give thoughts a greater significance than emotions. Deep emotions are not often recognized as the result of thoughts. Frequently, they are externalized as part of something else, for instance, assigning fear or evil to a target, though they are personal and subjective. David Bohm proposes that thought creates an image, and emotions do not regard it as a thought but as reality. Bohm sees this as the root cause of problems for individuals and humanity as a whole. A change in perspective is needed – a way to release and let go of the undesired. If you're familiar with Reiki, you can start the way you have learnt to begin a Reiki session and include the Reiki symbols you've learned, especially the symbol for treating emotions. If Reiki is new to you, follow the instructions while listening to your feelings.

1. Emotions are most often felt in the chest and head areas. Initially, place both hands on your chest or one on your forehead and the other on your chest.

2. Close your eyes. Naturally turn to listen more strongly to your inner self, to feelings and thoughts.
3. Breathe deeply. Breathing relaxes, the body gets oxygen, and tensions leave with the exhalation.
4. Imagine how, with each inhalation, an energy wave travels through your body from your feet toward your head. If it feels more natural the other way, you can do that too. With the exhalation, your attention moves back to your feet, and the wave passes over again.
5. With the gentle, healing wave, you can allow your body to relax and ask for all the qualities you need most at this moment – trust, love, clarity, patience. Imagine the wave as water-blue or whatever color and type that feels good right now.
6. Let the waves flow at their own pace, listening to them. Let them move and cleanse your energy body as needed; sometimes they may move faster, other times slower. Allow tears to come too – they also cleanse and release. You might identify critical points – where you can linger longer.
7. Finally, bring your hands together in front of your chest and end the exercise mentally by saying thank you.
8. If it feels right, you can now approach the subject or issue that triggered your reaction. Look at it with new eyes. Can you gain new insight, a new perspective? It's always best to go back to the source, to unravel or face the cause of the emotion. You can also write down or draw what you've experienced, what you feel right now. Sometimes, the exercise needs to be repeated—perhaps more processing time is needed.

Tears and even a good cry can be truly cleansing and liberating. In today's society, crying is often seen as a weakness. It's something to be hidden, as it has been customary to view it as a sign of weakness. However, crying is a connection between the heart and the head – emotional awareness, the sensitivity to understand and experience a situation on a deeper emotional level. Crying can be a pathway to your innermost self, to your soul. Things start to loosen, knots unravel.

In the old days, weepers played an important role in society. Crying has been a tradition among Baltic Finnic peoples, practiced in weddings and funerals through weeping or lamentation songs known to exist since the Iron Age. Like seers, weepers held an important place in the community. Lamentation songs were also an interaction with the beyond. By crying out loud, people spoke to the deceased, asking them to return to greet their loved ones, for example, as birds. There are also stories of the dead sending messages to the weepers through dreams. Thus, see crying as a strength and as forming a deeper connection.

Try Reiki's emotional healing for fear or anger too. Our emotions signal that something needs to be noticed and acknowledged. Fear and anger narrow our field of vision, our ability to see things correctly. Something is not aligned with your inner self, and you may not be able to envision a bright future. If such a situation arises, examine the basis of your views. Not everything is as it appears at first glance, so seek the truth. The best way forward is to take a step back and see the whole picture from a new perspective. Move from opposition toward unity. Also, accept that you don't need to change the whole world. It's enough to be in this moment and connect with your inner self. Finally, place your hands on your heart and send yourself love. Take care of yourself, both spiritually and physically, as these go hand in hand.

Timeline Healing

I have a dream of diving and war. In the first dream, I see light ahead, sunlight, but I am deep in dark blue water, and the journey to the surface is long. The dream feels good, even though there's no certainty of reaching the surface. The dream changes, the dark tone is still present, but the water is now fog, and it's not silent anymore; I hear the sounds of cannons and guns. I try to escape.

A few days later, I fall ill with COVID and am feverish for three days with peculiar muscle aches. The pains are on the outer sides of both legs – thus, almost at the seams – and extend all the way up to the upper part of the buttocks. The pains always start in the evening; I feel nothing during the day.

Only stretching helps, but trying to sleep in the yoga swan pose all night is, of course, challenging. Only that position seems to help and stretch enough.

My sister-in-law Maria is an emotional reflexology therapist. She told me that the nerve path at those points on the leg is the heart path, which relates to sorrow, deaths, separations, or being abandoned, while fear is on the muscle. She suggests tuning in to what trauma this could relate to.

I do timeline healing exercise because those war dreams seem to point to my grandparents, who lived during World War II. Maybe the war dreams also relate to the collective fear associated with the war in Ukraine that started just a couple of months earlier. In the healing, my intention is to get rid of the pain and relieve stress – whatever its cause. And of course, to send energy and love, right where it's needed most – into this moment, the past, and also the future!

1. Close your eyes and breathe deeply. If you have a specific area where you need energy, you can keep your palms there or on your heart.
2. Visualize in your mind walking along a timeline, a time spiral. It symbolizes your life path, all the phases that have brought you to this moment.
3. You don't need to think about the events of your life specifically. Just walk in your mind along the timeline and stop occasionally at "knot" or "pearl" points – moments that feel significant. Note the feelings and images at these points.
4. Send healing love especially to the knots. Store the energy of pearl points with you on the journey.
5. Do this as long as it feels right. Sometimes, even if you don't try, a moment pops into your mind, like suddenly remembering your childhood teddy bear and feeling what it's like to hold it.
6. Sometimes, you might feel another timeline, a time spiral, intersecting with yours – maybe another person's. You can also send energy to your shared journey and interaction.
7. Finally, return to the present and mentally thank yourself for the time travel.

As I do the timeline healing, I allow myself in my mind to go exactly where the healing is needed. I strongly feel chills in my leg, tears stream from my eyes. My maternal grandmother comes to my mind. When the war broke out, she was only 17 years old and had moved from the countryside to Helsinki a couple of years earlier. In my mind, I hear the sound of sirens, running, explosions. Fear, anxiety, sorrow. I switch legs. My thoughts go to my paternal grandfather, who was also just 17 when the Continuation War (1941) broke out. Tears flow, I strongly feel emotions in my stomach and legs. I can only express all this with these lines I wrote:

When time disappears around you and
only this moment remains,
when the burdens handed down
through generations fall away,
when your own past and future are meaningless,
when all roles disappear,
and only your core remains.

Tears are like droplets, rain. Suddenly, I realize I'm following water, through ages. The same water that was there during my grandparents' time, and their parents' before them, is still here after we're gone. Water unites and binds us together. I feel love and extend it to my grandparents. We are surrounded by it. Love frees them and gives me the strength to move forward. I also send energy to the future. The pain has disappeared and does not return.

"Water is a sacred force
of the thirsting heart,
in whose rhythm still
the stream of hope flows eternal."
– Aune Leskinen (From the poem *"The Eternal Stream of Our Hope,"* 1997)

This poem was written by my late maternal grandmother and caught my eye exactly on the 100th anniversary of her birth from a collection of poems she left me as a legacy. I was reminiscing and reading her poems that day.

Many of the poems are sad, full of longing and loss. She poured her feelings, experiences, and illnesses into her poems. Water often bubbles up in her poems. She writes about the stream of life, the beauty of ice crystals, and babbling spring brooks. As I flip through her poems, the timeline healing concludes when I read the following lines from one of her poems *"My Heart's Peace, a Necklace of Memories"*:

"I walk and string memories
like pearls on a necklace
from which I draw my strength,
feeling beneath my steps
the beating of my heart
radiating peace and light."

~

THE AGE OF AQUARIUS

After swimming again, I feel serene and calm. I avoid checking my phone because, even if one does not follow the news, stories about the economic situation, electricity prices, or the war in Ukraine pop up from different sources, while pandemic news has, at least momentarily, receded into the background. My great-grandmother's stories about the Age of Aquarius and end-times prophecies come to mind. I wonder, are we living through that now?

The Age of Aquarius is both an astronomical and astrological concept, determined by the position of the Sun in relation to the constellations: the Sun moves from the zodiac sign of Pisces into Aquarius. The Age of Aquarius is actually a concept associated with a new era and way of thinking. The transition is interpreted, especially in new-age circles, as the beginning of a new epoch and a turning point in human history, which is thought to include a shift to post-materialism, moving beyond materialistic thinking. People understand their deepest, true essence.

"The Age of Aquarius demands that we hold a worldview
that none of us is separate from

one another nor from the planet."
– Andrew Lutts, *How to Live a Magnificent Life: Becoming the Living Expression of Higher Consciousness*

However, the concept always brings me a sense of fear and tension. I first heard about the Age of Aquarius and the predictions of the end of the world in my teenage years from my great-grandmother. Often, when we went to see her, she would be listening to priests' sermons and other church recordings on her cassette player: The end of the world was coming. I read about the *Book of Revelation* and the millennium kingdom, the antichrist, floods and earthquakes, locusts, deadly diseases, wars, the blood-red moon, and how only believers would be saved from all this. According to my great-grandmother, we were already living in the Age of Aquarius – this was all happening back in the 1980s.

I must admit that I have largely overlooked astrology in the past. Probably because astronomy is science and relates to physics, but astrology is considered more of an entertainment, without a real basis, although, for example, Hindu astrology (Jyotisha) or Indian astrology, now often referred to as Vedic astrology, are systems of knowledge thousands of years old. Jyotisha, which means light (of the sun, moon, or celestial bodies), in texts combines astronomy, astrology, and the science of time, utilizing the movements of astronomical bodies.

One reason I have overlooked astrology is also that I haven't liked the idea that all people born at a certain time of the year would be just a certain way and alike. Astrology has not been able to answer, for example, one question I consider essential: how can twins born a minute or two apart be different? Or how can children born in the same hospital at the same time be so different, even though the planets are exactly in the same position? However, Vedic astrology should be able to provide an answer to this. It distinguishes what a person truly is, what they want to be, what they think they are, and what others think they are.

In recent years, however, I've noticed that my image of astrology doesn't reflect the truth or reality. Perhaps my view on this matter has been limited, and I should explore it more closely, especially since the Age of Aquarius is indeed also an astronomical term? Moreover, as I've often stated, we are not

Figure 24. Precession. One complete rotation back to its current position takes about 25,725 years. About halfway through the rotation, the rotation axis points toward Vega, the main star of the Lyra constellation.

separate but one and the same energy and consciousness with our entire universe. Astrology is similar to reading Tarot cards, a kind of synchronism. Astrology can be seen as a "key" or "guide" to learning to understand this reality. A new language, so to speak, to interpret what we experience.

Astrological ages are based on the 12 signs of the zodiac and the position of the sun in these signs as observed from Earth. The ages move in the opposite direction compared to zodiac signs, so the Age of Aquarius follows the Piscean age. During the Age of Aquarius, the sun is in the constellation of

Aquarius at the time of the spring equinox. The periods are a result of the Earth's precession, or the rotation of the tilt of its axis (Fig. 24). A complete rotation takes about 25,725 years. According to the Egyptians, everything renews itself with each "great year" cycle. By this, they meant the rotation of the axis, which is easy to understand by following, for example, a spinning top. After a great year, at the time of the spring equinox, the stars are essentially in the same position, in the same relationship to Earth again. During that time, all 12 zodiac constellations are traversed. Each period lasts about 2,150 years, which corresponds to 30 degrees on the zodiac.

In Western astrology, so-called *tropical* astrology is used, where the zodiac signs are locked in position from 2,000 years ago. At that time, the zodiac signs were in the same position as the constellations of the same name and lived, or still live, in the Age of Pisces. The constellations are always the same distance from the spring equinox point, for example, the Sun is in the sign of Aquarius always around January 20–February 19. In Indian *sidereal* astrology, the same-named signs and constellations correspond to each other, following the actual situation of the starry sky. It is said to be more accurate and stable. Therefore, in tropical astrology, time is calculated according to the equinoxes and solstices (two of each per year), while in sidereal astrology, it is according to the stars. Naturally, these are just different interpretations, which a skilled astrologer can take into account.

According to astrologers and astronomers, transitions from one age to another do not happen suddenly but gradually, without any precise boundary. Incidentally, a one-degree shift on the zodiac corresponds to about 72 years, which is nearly the length of the current average life expectancy. Thus, astrologers say we have lived in the Age of Pisces, and now we are in the transitional phase to the Age of Aquarius. This topic was widely discussed after World War II in the 20th century, and a new age (New Age) was said to begin. For instance, in the 1960s musical Hair, they sang about the beginning of the age of Aquarius and the true liberation of the mind: *"This is the dawning of the age of Aquarius."*

Vedic astrology is also referred to as the "eyes" of the Vedas, as it allows one to see life experiences more clearly and also make predictions about the future. Vedic astrology is thousands of years old and is fundamentally based on the idea that everything is interconnected. The movements of the planets are not separate from what happens to us. In fact, it's even said that by understanding astrology, you also understand the secrets of the universe.

I learned the basics of Vedic astrology in a course taught by Joni Patry, an American Vedic astrologer with over 40 years of experience in the field. Both in Vedic astrology and in Western astrology, there is a certain formula for interpreting events. The combined influence of planets is significant. The transitions are also meaningful – that is, how the planets move in the sky relative to the birth chart. Different events, or any significant moment for oneself, are thus examined by considering transitions, planetary relationships, and their combined effects which, together, tell about different aspects of life. Star charts also revealed why a book on water is important and timely for me right now.

I studied the charts in April 2022. From about a year forward, Jupiter was in the constellation Pisces, and at the same time, Neptune was in the Sun, meaning that water is everywhere. Jupiter and Neptune are planets associated with water, and Pisces is also a water element sign. Their combined influence means a lot of water-related events: floods, storms, rain, big upheavals. I also studied my own chart, and in it, Jupiter plays an important role, as the Sun activates Jupiter in the third house. In astrology, houses represent different areas of life, and the third house is related to communication, thinking, learning, expression, and also writing. Water is also prominently present in my own chart; my rising sign in Vedic astrology is Scorpio, a water sign, and my Sun sign is Cancer, which is also a water sign. So it's no wonder I'm writing this water-themed book.

Roughly speaking, the Age of Pisces lasts from about 100 BCE to 2100. Due to the inaccuracies described above, it's impossible to pinpoint the exact year, and it's not even important, as things often develop gradually, and the world changes slowly. In any case, we are living right now in this time of change, both astronomically and from an astrological perspective.

Apocalyptic views of the Age of Aquarius are not the desired growth related to the new age, but depending on the perspective from which the Age of Aquarius is viewed, it can appear as a threat to the current state of humanity. The Age of Pisces is considered a state of balance, associated with religiosity, hierarchies, political ideologies, power mechanisms, and Christian values prevailing and guiding. The Age of Aquarius is especially talked about in religious circles as a period associated with the second coming of Christ in the midst of the end times, saving the believers. This passage does not specifically emphasize apocalypse or the end of the world but rather focuses on collective salvation and a new kingdom, achieved gradually.

According to astrology, the Age of Aquarius is associated with spiritual growth. More and more people understand their mission and their place – and that we are not separate but together part of something greater. The notion of the growth of consciousness and ascending to the next level can be seen precisely as part of the transition to the Age of Aquarius. We are more ready to question and not take information for granted but go to the source of knowledge seeking the truth. Similarly, we are moving more away from individualism toward thinking about the entire community, from pursuing self-interest toward a collective vision. We begin to expand our view from Earth to examine the whole universe and the purpose of everything. Finding one's place is a part of everyone's life, but increasingly, and perhaps more strongly than ever, becoming whole and expanding consciousness is too.

Often, the coming of a new age is thought to be accompanied by the arrival of a new teacher, leader, guru who guides and instructs in growth. However, during the Age of Aquarius, teaching seems to involve one trait that is well summarized in an article in Ultra magazine (Ultra 9/2010, 12) that I noticed fitting this perfectly: *"Characteristics of the new age person include staying in the background, such as leading without a leader, teaching without a teacher, and helping without knowing who helps. Does it sound like Zen? No, but rather like the Age of Aquarius!"*

The true teacher's task is not to elevate themselves above others but to teach their students to be their own gurus, to find the answers within themselves.

Pekka Ervasti, a pioneer of Theosophy in Finland at the turn of the 19th and 20th centuries, states about the Age of Aquarius that it will bring out both the worst and the best in people. It's typical for us to want to cling to the familiar, safe, and proven, and change always involves resistance. The potential for spiritual growth exists, but unfortunately, the path to new understanding often occurs through purgatory. Keeping direction amidst suffering is crucial. Pekka Ervasti actually believed that the transition to the Age of Aquarius happened at the beginning of the 20th century.

Thus, what should be kept in mind, whether we live in the Age of Aquarius or not? Definitely a spirit of change. Everyone should teach themselves to manage and accept changes, to understand what happens in change. Change always involves certain stages. One of these is resistance. We resist change and idolize the past, wanting to return to it before coming to a phase where the focus increasingly turns to the future. We start to see the good aspects of change and seek our place in the new situation. It's also common for some to still idolize and dream of the good old times while others have already turned their gaze to the new. So, prepare for changes and learn to face them. This involves finding the most personal and suitable ways to listen to oneself, to turn inward, seek answers, and perhaps even explore one's own spirituality.

In my view, the era of New Age spirituality as it has been since the wars, is over. Or at least it's time to move forward from the concept of spirituality that many have held. The right path is not just to turn inward and passively accept everything with gratitude. It's also necessary to act based on fundamental values, such as love, kindness, and justice. Thoughts and feelings matter, and it's important to find a connection to one's inner self, for example, through meditation. All this should also be brought into everyday reality and practice. It's not enough to acknowledge the situation, like the desire to live from love, with love as the foundation and guideline, if after the practice, you return to shouting at your spouse and children. It should be possible to live according to inner peace and fundamental values, to find solutions to challenges, living as one thinks.

Perhaps the Age of Aquarius is just another reminder of how we should live? I found in my great-grandmother's Bible (from 1953) the Great Commandment of Love underlined. Next to it, a big exclamation mark was drawn for: *"And you shall love the Lord, your God, with all your heart, and with all your soul, and with all your mind, and with all your strength. Love your neighbor as yourself. There is no commandment greater than these"* (Mark 12:30–31).

In *Hyperreality*, I presented five principles to which modern physics guides us: Oneness, Consciousness, Interaction, Complementarity, and Freedom. All these together lead to the most important: the Principle of Love – that we should live from our hearts. This could be summarized as: *"The most important thing always is love."*

I did a meditation for the Age of Aquarius, saying: *"Leave behind all that is extra, unnecessary, all that does not serve your soul's mission right now. It's time to step forward, dive into new waters, unknown, but trusting that you are not alone. We do this together. You are guided toward your true purpose."*

EPILOGUE AT THE FOUNTAIN OF EVERYTHING

"Your job on Earth is not to learn,
but to remember Who You Are.
It is your sole purpose.
That is, your soul purpose."
– Neale Donald Walsh

~

In a Florentine document, Leonardo da Vinci was described as the "Master of Water," and not without reason. Throughout his life, da Vinci studied water and made several inventions related to it: among others, a prototype of a submarine, water-lifting devices, dams, and bridges. Today, he is also considered one of the forefathers of hydrology, the science of water, as he studied water's occurrence, properties, and cycle. Water also appears in most of da Vinci's paintings, and in his last years, he drew many studies attempting to understand water's movement and its nature. To describe moving water alone, da Vinci had 67 words. His view was that you cannot control or understand something unless you can describe or measure it. Sometimes his drawings depict flooding water and even catastrophes. He experienced the devastating flooding of the Arno River, which swept away people, animals, and houses when he was 14 years old. Perhaps by trying to understand water and control its movements with his devices, he was also dealing with his own fears? Da Vinci thought that by understanding how water behaves in the air and on land, he could unravel the mystery of life through water.

In many spiritual traditions, the journey to becoming a master of esoteric or hidden knowledge goes through certain stages. One seeks answers to mysteries – knowledge that we already know deep inside, as if we were bringing it back to our mind and memory. Fundamentally, however, the spiritual journey is a search for an answer to the question: Who am I, really?

During the journey, you encounter your own limitations on different levels, your fears, and surrender to being exactly who you are meant to be. After relinquishing your burdens, including your fears, it's time to arrive at the source, the core of all knowledge, and join other masters.

The master's task is to understand, pass on, and also activate others to control the learned knowledge. All this has been done in the past in mystery schools, whose task was also to protect sacred teachings and rituals.

Knowledge of water cannot remain secret. Water now needs masters who will act on behalf of water in various ways.

Rumi, a Persian poet and mystic who lived in the 13th century, says: *"You are not a drop in the ocean. You are the entire ocean in a drop."* I have heard this quote many times but now I really understand it deeply.

By changing our own way of looking at reality and our perception, we simultaneously change the world as well. Even if you may feel like an insignificant little drop, you can be as vast in significance as the sea.

The Message of Water

- Water is sensitive. It changes through interaction and stores, transmits, and amplifies information related to everything it comes into contact with.
- Water can be influenced in different ways. Light, sound, and everything that comes into contact with water affects it.
- We impact water with our consciousness and also unconsciously. The water is not the same after encountering you as it was before.
- Waters differ. Spring water from a mountain stream (which is mineral-rich, structured, "living" water) is the best for drinking, but it's possible to influence all water.
- Especially water that has been in sunlight and just melted contains structured water, the same as in our cells. The body also converts ordinary water into this form.
- Thoughts and emotions also affect water. Remember, we are almost entirely made of water. What applies to water likely applies to us as well.
- By protecting water, we protect the Earth and ourselves. Each of us has an impact.

• Seawater grounds the best, that is, brings negative ions into the body, which may be lacking in the body during inflammatory conditions.

• Make choices that are best for water – it's also for your own good.

• Learn to understand the messages from the unconscious mind. Your innermost being's current issues must be dealt with in your everyday consciousness. They can manifest in different ways, for example, as synchronicities or dreams.

• Form an experiential connection with water or strengthen an existing one. Challenge yourself by doing something new. By strengthening this connection, you also connect more strongly to the all-encompassing collective field of knowledge.

• Take inspiration from water and live accordingly: When it rains outside, let the water wash away sorrows and excess burdens. When white clouds drift across the sky, remember to dream. When unique and delicate ice crystals form on the window, be thankful for the precious moments in your life. Water brings flow to your life, living in the moment, letting go, joy, light, and play!

Giving Thanks to Water

Thank you, water, timeless and ageless.
Thank you for blessing, cleansing, nourishing,
and healing me.
Thank you for having done so, for doing so now,
and also in the future.
Thank you for teaching me by providing timeless information and telling
your eternal story as we touch you with our lips and drink.
So that we may be able to support your sanctity and purity in us all and in
the sources of water everywhere on the planet,
may you run free, pure, unobstructed.

Give us purity of heart and mind, when your waves touch us.
Thank you.

ABOUT THE AUTHOR

Johanna Blomqvist, Ph.D., is a physicist, entrepreneur, and internationally published author. Blomqvist is a cold-water swimming enthusiast and approaches any subject with an open and inquisitive researcher's attitude.

In her work, Blomqvist uniquely combines many different disciplines and methods, often approaching the frontiers of science as a pioneer. Blomqvist's previous works include *From Quantum Physics to Energy Healing – A Physicist's Journey to Mind and Healing* (Mindstream Publishing, 2018) and Hyperreality – *Beyond the Horizon Where Physics Meets Consciousness* (Mindstream Publishing, 2021).

More information about the meditations mentioned in this book, as well as *The Messenger of Water™* online course for deepening your connection with water and becoming it's messenger can be found at www.johannablomqvist.com

GLOSSARY

Absorb – capture, collect

Amorphous – non-crystalline, shapeless. An amorphous substance is a solid that is not crystalline, meaning it is not organized in a lattice. The most well-known amorphous material is glass. Plastics, for instance, can consist of polymers that are in an amorphous state.

Anomaly, anomalous – unusual, abnormal, irregular. Something that a general rule does not predict.

Avogadro's limit – or Avogadro-Loschmidt limit. The concentration limit in a diluted solution below which it is improbable that any of the original molecules remain in the solution.

Capillary action – the tendency of a liquid to rise in narrow tubes or to be drawn into small openings against external forces, such as gravity. This action is due to the water and glass's attraction, adhesion, being as strong as the water's internal cohesion. Water rises from a plant's roots toward its top primarily because of capillary action.

Cataclysm – a large-scale upheaval, comes from the Greek word kataklysmós, meaning flood.

Causal – obeying a cause-and-effect relationship

Colloid – from the Greek kollodes, meaning glue-like or sticky. A mixture that is an intermediate between homogeneous and heterogeneous mixtures, where particles are finely dispersed but larger than those in a homogeneous mixture.

Configuration – spatial structure

Concentration – the amount of a dissolved substance in a solution

Contamination – polluted, contaminated, poisoned. The presence of an unwanted component in a liquid, material, or environment.

Covalent bond – a chemical bond formed between two elements (such as hydrogen and oxygen). In this bond, electrons are shared between these elements.

Delocalization – an electron does not belong to any specific bond or atom but collectively to the entire molecule.

Electric dipole – a pair consisting of two equal but opposite charges.

Entanglement – a phenomenon in quantum physics proposed by E. Schrödinger in the 1920s and experimentally proven for the first time in the 1970s–80s. In entanglement, two particles can be separated by large distances, yet a change in one particle's state, such as its spin, will instantaneously affect the other.

EZ , Exclusion zone – a term by Gerald Pollack referring to the area formed by the effect of hydrophilic surfaces on water, which repels foreign atoms and molecules.

Free electron pair, or unbound electron pair – an electron pair on the outer shell of an atom in a compound that is not involved in bond formation. Free electron pairs still play a role in the compound's structure and properties.

Hydrophilic – attracting water

Hydrophobic – repelling water

IR, infrared radiation – electromagnetic radiation with a wavelength longer than visible light but shorter than microwaves.

Modulate – to adapt, affect, adjust. In music, to change the key.

Morphic fields – a theory developed by biologist Rupert Sheldrake suggesting that nature consists of wholes with inherent systemic fields, or morphic fields. More on this can be read in my book *Hyperreality*.

Mystery – enigma, unknown

Myth – a story, legend, or belief, possibly imagined or without a rational explanation

Nafion – a copolymer consisting of tetrafluoroethylene and perfluoro-3,6-dioxa-4-methyl-7-octene-sulfonic acid. Nafion is a conductive polymer and has been extensively researched in the fuel cell industry.

Non-local – happening from a distance

PEAR – Princeton Engineering Anomalies Research. A laboratory at Princeton University founded by Robert Jahn (project director) and Brenda Dunne (laboratory engineer). PEAR investigated, among other topics, remote viewing, human-machine interaction, and the general effect of the mind on the material world.

pH – indicates the acidity or alkalinity of a liquid. Pure, neutral water is 7 on the pH scale. If the pH is below 7, the water is acidic, meaning there are more H^+ ions in the water, and if above 7, it is alkaline, meaning a higher OH^- concentration.

Polar molecule – a molecule's charge is unevenly distributed. The water molecule is polar because hydrogen has a strong positive charge, and oxygen has a negative charge.
Polymer – a molecule where several molecules, or monomers, are chemically bonded together.
Probiotics – live microbes that strengthen the gastrointestinal flora.
Quantization – the phenomenon where, instead of a continuous distribution, only certain values or states are possible, for example, energy can only take certain values instead of a continuous distribution.
Quantum coherence – Two waves are in complete coherence when they have the same frequency and wavelength. In quantum coherence, a particle's waveform splits into two, and these waves interfere with each other.
Quantum mechanics – Often used interchangeably with "quantum physics." A theory describing the behavior of atoms and particles.
Reductionism – The higher-level whole consists of lower-level parts and can thus be broken down into smaller components. Opposite of holism.
Spin – A property of elementary particles, an intrinsic angular momentum, resulting from the particle's spin.
Stability – not easily losing balance.
Supercooling – A liquid cools below its freezing point without becoming solid. Water needs a freezing nucleus to freeze, such as a dust particle or an ice crystal. Without these, water can remain liquid in the lab even at -42 °C (-43.6°F).
UV, ultraviolet radiation – Electromagnetic radiation with a wavelength shorter than visible light but longer than X-rays.

ACKNOWLEDGMENTS

Writing this book has been an intriguing journey of exploration – truly a dive into our deepest essence. I have waded through hundreds of studies and books about water, encompassing thousands of years of work to understand it. After my previous book on reality, I imagined lightly plunging into water, but this aquatic journey turned out to be deeper than I could have anticipated. The dive has been cleansing, opening up entirely new dimensions for me. Thank you, water, for this journey and guidance!

I want to thank the publisher of my Finnish book Viisas Elämä Oy, for their trust — thank you for publishing my book. Thanks to the entire publishing team and especially Harri Nyman, Julia Alakulju, Henna Helmenmaa, Enna Kulo, Jonna Nisu, Sanna Vaara. I have been able to let go of the book knowing it is in good hands. Warm thanks also to Sanna, Henna, and Julia for your books, wonderful stories! Thanks also to my fellow author Mervi Enqvist for the peer support. It's fantastic that our books were being released at the same time. Much success to your book, also internationally! Special thanks also to Saara Metsäranta, who edited the Finnish book.

Thank you, Tarja Kallio-Tamminen, Ph.D., for your valuable comments on the book. I greatly appreciate your insights.

Thanks to Marjut Saariniemi for the hypnosis journey, my sister-in-law Maria Blomqvist for the emotional reflexology therapy consultation, and Elena Gushchina for the water story. Intriguing and eye-opening experiences, all of which helped deepen my understanding of water and myself!

Thanks to Maria Nordin for the water conversations and fantastic experiential story! It's great that water has such a messenger as you!

Thanks also to my autumn 2022 Reiki teacher course students for the help with water experiments and fantastic discussions between earth and sky, and sometimes beyond.

I am grateful to Amy Thomas, Ph.D., for editing the English version of this book. Thanks to Rosie Noble for valuable comments on the book.

Thanks to my mother Sinikka and my sister Katri for their help with the background work. You are an important part of my journey. Thank you for your support. Special thanks also to Katri for proofreading and final editing of the book.

My greatest thanks go to my family and especially my spouse, Anssi. Thank you for supporting and understanding me during the writing of this book as well! Thank you, Anssi, for your clarifying comments on the book – I cannot express in words how much I value your expertise, your insights, and the fact that you sometimes read my book even at night. You make me complete. <3

Thank you, Teo and Stella, for the rewarding conversations, swimming trips in sometimes ice-cold water, and generally for being in my life.

REFERENCES

Books:

M. Aurelius, *Meditations of Marcus Aurelius*, translation: Jeremy Collier. London: Walter Scott, 1887.

S. Bartlett, *Working with the Tarot - Unlock the Secrets of Your Destiny with the Tarot,* Octopus Publ. Group, 2010.

F. Batmanghelidji, *Your Body's Many Cries for Water*, Global Health Solutions, Inc.; Third Edition, 2008.

S. de Beauvoir, *The Second Sex*, part II, s. 744, translated: Constance Borde and Sheila Malovany-Chevallier. First Vintage Books Ed., May 2011. Le deuxieme sexe, Editions Gallimard, 1949.

Hildegard von Bingen, *Causes and Cures*, (the original is from 12th century), translation: Priscilla Throop. MedievalMS, 2012.

W. Bengston, *Hands on healing: a training course in the energy cure (CD audio set)*. Sounds True; 2010.

J. Blomqvist, *Hyperreality – Beyond the Horizon where Physics Meets Consciousness*, Mindstream Publishing, 2021.

J. Blomqvist, *From Quantum Physics to Energy Healing – A Physicist's Journey to Mind and Healing*, Mindstream Publishing, 2018.

D. Bohm, *Wholeness and the Implicate Order*, London, 1983.

D. Bohm, M. Edwards, *Changing consciousness - Exploring the Hidden Source of the Social, Political, and Environmental Crises Facing Our World,* HarperCollins, 1991.

N. Brendborg, *Jellyfish Age Backwards: Nature's Secrets to Longevity,* Hodder & Stoughton.

M. Chaplin, *Structure and properties of water in its various states,* Encyclopedia of Water: Science, Technology, and Society, Ed. P. A. Maurice, Wiley, 2019.

P. Chappell, D. Andrews, *Healing with Homeopathy,* Gill & Macmillan Ltd, 1996.

B. Dunne ja R. Jahn, *Being & Biology - Is Consciousness the Life Force?*, ICRL Press, 2017.

M. Emoto, *The Miracle of Water*, ATRIA Paperback, Beyond Words Publ. Inc., and Simon and Schuster, Inc., 2007, translation: David A. Thayne.

S. Freud, *The Interpretation of Dreams* (Die Traumdeutung, 1900), Flame Tree Collections; Deluxe edition, 2020.

H. Helmenmaa, *Metsän hoitava syli - Rohkeasti omalla polulla*, Viisas Elämä, 2022.

Hippocrates, *On Airs, Waters, and Places, (the original version circa 400 BC),* Kessinger Publishing, LLC, 2004.

M. Hjelt, *Maaginen vuosi*, SKS Kirjat, 2020.

M.-W. Ho, *The Rainbow and the Worm: The Physics of Organism,* World Scientific Publ., 2008.

C.G. Jung, *In Archetypes and the Collective Unconscious* (Über die Archetypen des kollektiven Unbewussten,1934), 2001.

C.G. Jung, *The Red Book,* the original version Wink Books, 1930, new version W. W. Norton & Company, 2009.

T. Kesäläinen, A. Kejonen, *Pyhien vesien äärellä*, Salakirjat, 2022.

H. Kronberger, S. Lattacher, *On the Track of Water's Secret - from Victor Schauberger to Johann Grander*, Uranus, 1995.

B. Kröplin, C. R.C. Henschel, *Water and its memory: New astonishing insights in water research*, AT Verlag; 1. Auflage, 2017.

M. Lehtimäki (ed.), H. Meretoja, A. Rosenholm, *Veteen kirjoitettu - Veden merkitykset kirjallisuudessa*, SKS, Helsinki, 2018.

D.S. Lihatšov (Finnish translation: M.-L. Jaakkola), *Nestorin kronikka*, WSOY, 1994.

V. Linna, *Sininen Sinfonia - Oodi vedelle*, Basam Books, 2022.

T. Maunonen, *Suuri Unikirja*, Gummerus, 2011

L. Märcz, *Feral children. Questioning the human-animal boundary from an anthropological perspective,* Working Papers of the Department of Anthropology and African Studies of the Johannes Gutenberg University Mainz, 179, 2018.

L. McTaggart, *The Intention Experiment: Using Your Thoughts to Change Your Life and the World,* Frcc Prcss, 2007.

L. McTaggart, *The Power of Eight - Harnessing the Miraculous Energies of a Small Group to Heal Others, Your Life and The World, Atria Books, 2017.*

C. Ober, S.T. Sinatra, M. Zucker, *Earthing - The most important health discovery ever!* Basic Health Publications, Inc.; 1st edition, 2010.

L. Pauling, *In Hydrogen Bonding*; Hadži, D., Ed.; Pergamon Press: New York, 1959.

Plato, *Timaeus and Critias*. Penguin Classics; Revised edition, 2008.

H. Peltola, *Whole and One: Seeing and being the World,* Nostetuotanto, 2020.

S. Perko, *The Homeopathic Treatment of Influenza: Surviving Influenza Epidemics and Pandemics Past, Present, and Future with Homeopathy*, Benchmark Homeopathic Publications, 1999.

M. Planck, *Where is Science Going?*, New York : AMS Press, 1932.

G. Pollack, *The Fourth Phase of Water - Beyond Solid Liquid Vapor*, Ebner and Sons Publishers, 2013.

G. Pollack, *Cells, Gels and the Engines of Life - A New, Unifying Approach to Cell Function*, Ebner and Sons, 2001.

A. Schmemann, *Off Water and the Spirit,* St. Vladimir's Seminary Press, 1974.

K.A Sharp, *Water: Structure and Properties,* Encyclopedia of Life Sciences, John Wiley & Sons, 2001.

R. Sheldrake, *Science and Spiritual Practices, Counterpoint LLC, 2019.*

J.H. Smart, *The Cold Therapy Code: Rediscover Your Vitality Through Cold Exposure - The 3 Simple Cryotherapy Methods for Reducing Stress, Improving Sleep, and Increasing Energy,* Roger Davies, 2021.

S. Vaara, *Mökkihulluutta ja mielenrauhaa*, Viisas Elämä 2022.

J. Vitale, I.H. Len, *Zero Limits: The Secret Hawaiian System for Wealth, Health, Peace, and More,* Wiley, 2008.

W. Hof, E. Epel, *The Wim Hof Method: Activate Your Full Human Potential*, Sounds True, 2020.

Yogi Ramachakara, *Water Cure: The Hindu Yogi System of Practical Water Cure, CreateSpace Independent Publishing Platform,* 2010.

Articles:

B. Alsabti, D. Radha, F.M. Dashti, D. Sabarathinam , *Fluoride in Groundwater*, In book: *Hydrogeochemistry of Aquatic Ecosystems*, Nov. 2022.

A. Ansaloni, *Effect of Lourdes water on water pH,* Bollettino Chimico Farmaceutico 142(5): 202-5, 2003.

Bengston's studies:

B. Bengston, M. Nies, *Energized Water Study*, https://bengstonresearch.com/content_assets/docs/Energized_Water_Report.pdf

B. Bengston, M. Nies, T. den Ouden, C. van Veenendaal, *A Replication of a Clinical Social Experiment of Device-infused Resonant Water*, https://bengstonresearch.com/content_assets/docs/Replication-Clinical-Social-Experiment-Resonant-Water.pdf

W. Bengston, M. Moga, *Anomalous DC Magnetic Field Activity during a Bioenergy Healing Experiment,"* Journal of Scientific Exploration, Vol. 24(3), Fall 2010.

W. Bengston, M. Nies, *A Double Blind, Placebo Controlled Clinical Trial on Hospitalized Covid Patients Using Informed Water*, J Sci Explor. Vol. 37(1), Spring 2023.

W. Bengston D. Krinsley, *The effect of the laying-on of hands on transplanted breast cancer in mice.*, J Sci Explor.;14(3):353–364, 2000.

W.F. Bengston, M. Moga, Resonance, placebo effects, and type II errors: some implications from healing research for experimental methods, J Altern and Complement Med. 13(3): 317–327, 2007.

L.J. Lerner, A. Bianchi, M. Dzelzkains, *Effect of hydroxy-urea on growth of a transplantable mouse mammary adenocarcinoma.* Cancer Res. 1966;26(11):2297–2300.

S. Beseme, W. Bengston, D. Radin, M. Turner, and J. McMichael, *Transcriptional Changes in Cancer Cells Induced by Exposure to a Healing Method, Dose Response, July 2018.*

Benveniste's studies:

Benveniste under Review, https://www.nature.com/articles/340089b0.pdf

Y. Thomas, M. Schiff, L.Belkadi, P.Jurgens, L.Kahhak, J.Benveniste, *Activation of human neutrophils by electronically transmitted phorbol–myristate acetate*, Medical Hypotheses, Volume 54, Issue 1, January 2000, Pages 33-39

J. Maddox, J. Randi & W.W. Stewart, "High-dilution" experiments a delusion, *Nature*

volume 334, pages, 287–290, 1988. https://www.nature.com/articles/334287a0

B. Poitevin, E. Davenas, J. Benveniste, *In vitro immunological degranulation of human basophils is modulated by Lung histamine and Apismellifica,* Br. J. clin. Pharmac., 25, 439-444, 1988.

B. Chai, H. Yoo, G.H. Pollack, *Effect of radiant energy on near-surface water,* J. Phys. Chem. B. Oct 22;113(42):13953-8, 2009.

I.T. Cousins, J.H. Johansson, M.E. Salter, B. Sha, M. Scheringe, *Outside the Safe Operating Space of a New Planetary Boundary for Per- and Polyfluoroalkyl Substances (PFAS), Environ. Sci. Technol.* 2022, 56, 16, 11172–11179.

E. Del Giudice ym., *Coherent structures in liquid water close to hydrophilic surfaces,* J. Phys.: Conf. Ser. 442 012028, 2013.

E. Del Giudice, V. Voeikov, A. Lomonosov, A. Tedeschi, *The widespread properties of EZ water.* - A chapter in the textbook *Fields of the Cell: The origin and the special role of coherent water in living systems*, Research Signpost, 2015.

D. Deming, *The Aqueducts and Water Supply of Ancient Rome,* Ground Water, Jan-Feb; 58(1): 152–161, 2020.

R. Descartes, *Discourse on the Method* (the original 1637), translated in English Laurence J. Lafleur, New York: The Liberal Arts Press, 1960.

U. di Corpo, A. Vannini, *Retrocausality: experiments and theory,* www.sintropia.it, 2010.

G. Dhillon et al., *Hydrogen Water: Extra Healthy or a Hoax?-A Systematic Review,* Int. J. Mol. Sci., Jan 12;25(2):973, 2024.

B. Dunne, R.D. Nelson, R.G. Jahn, *Operator-Related Anomalies in a Random Mechanical Cascade,* J. Scientific Exploration, 2, No.2, pp.155-179, 1988.

D.C. Elton ym., *Exclusion Zone Phenomena in Water—A Critical Review of Experimental Findings and Theories,* Int. J. Mol. Sci. 21(14), 5041, 2020.

L. Fantappiè, Sull'interpretazione dei potenziali anticipati della meccanica ondulatoria e su un principio di finalità che ne discende. Rend. Acc. D'Italia, n. 7, vol 4, 1942.

A. Foletti, M. Ledda, E. D'Emilia, S. Grimaldi, A. Lisi, *Differentiation of human LAN-5 neuroblastoma cells induced by extremely low frequency electronically transmitted retinoic acid,* J. Altern. Complement. Med. 17: 701-704, 2011.

A. Foletti, M. Ledda, E. D'Emilia, S. Grimaldi, A. Lisi, *Experimental finding on the electromagnetic information transfer of specific molecular signals mediated through the aqueous system on two human cellular models,* J Altern. Complement. Med. 2012; 18: 258-261.

D. Florea, S. Musa, J.M.R. Huyghe, H.M. Wyss, *Long-range repulsion of colloids driven by ion exchange and diffusiophoresis,* Proc. Natl. Acad. Sci., 111, 6554–6559, USA, 2014.

B. Grad, *A telekinetic effect on plant growth,* Int. J. Parapsychol.,5(2):117–133, 1963.

B. Grad, *A telekinetic effect on plant growth: II. Experiments involving treatment of saline in stoppered bottles,* Int J Parapsychol. 6(4):473–498, 1964.

W.B. Hardy, *Collected Scientific Papers of Sir William Bate Hardy,* Cambridge University Press, 1936.

J.C. Henniker, *The depth of the surface zone of a liquid,* Rev. Mod. Physics. 21(2): 322-341, 1949.

M.-W. Ho, *Crystal clear, messages from water,* Science in Society 15, 24-25, 2002.

C.I. Honniball, P.G. Lucey, S. Li ym., *Molecular water detected on the sunlit Moon by SOFIA,* Nat Astros 5, 121–127, 2021.

S.T. Hsieh, G.V. Lauder, *Running on water: Three-dimensional force generation by basilisk lizards,* PNAS, 101 (48) 16784-16788, 2004.

Homeopathy:

I.R. Bell, G.E. Schwartz, J. Frye, B. Sarter, L.J. Standish, *Extending the Adaptive Network Nanomedicine Model for Homeopathic Medicines: Nanostructures as Salient Cell Danger Signals for Adaptation,* Nanoscience & Technology: Open Access

https://symbiosisonlinepublishing.com/nanoscience-technology/nanoscience-technology24.php

Iris R. Bell, MD PhD (Oct 2018), Pursuing the Biological Basis of Homeopathic Remedy Responses - References https://magicpillsmovie.com/uploads/attachments/cl1uv890302t1qih2r5ylu7t0-homeopathicresearchreferencesoct2018.pdf

J.R. Pereira Guedes , L.V. Bonamin , V.L. Capelozzi, *Water-Related Mechanisms Proposed for Storing and Transmitting Homeopathic Information:*

Putative Links with Biological Responses, Homeopathy, 107(03): 172-180, 2018.

A. Manzalini, B. Galeazzi, *Explaining Homeopathy with Quantum Electrodynamics, Homeopathy* 108(03): 169-176, 2019.

Inez Hua's (Purdue University) studies mentioned in Harald Franzen's article: Cleaning Water with Ultrasound, Scientific American, 23.4.2001. https://www.scientificamerican.com/article/cleaning-water-with-ultra/

K. Hyeon-Deuk, K. Ando, *Quantum effects of hydrogen atoms on the dynamical rearrangement of hydrogen-bond networks in liquid water,* The Journal of Chemical Physics 132(16):164507, 2010.

E. D. Isaacs, A. Shukla, P. M. Platzman, D. R. Hamann, B. Barbiellini, and C. A. Tulk, *Covalency of the Hydrogen Bond in Ice: A Direct X-Ray Measurement,* Phys. Rev. Lett. 82, 600, 1999.

R.G. Jahn, B.J. Dunne, *The Science of the Subjective*, Journal of Scientific Exploration, Vol. 11, No. 2, pp. 201–224, 1997.

W.B. Jonas J.A., Ives, F. Rollwagen, D.W. Denman, K. Hintz, M. Hammer, C. Crawford, K. Henry, *Can specific biological signals be digitized?* FASEB J, 20:23-28, 2006.

B.D. Josephson, *Beyond the 'theory of everything' paradigm: synergetic patterns and the order of the natural world,* June 2021, https://www.researchgate.net/publication/352368019_Beyond_the_'theory_of_everything'_paradigm_synergetic_patterns_and_the_order_of_the_natural_world

S. Kernbach, V. Zamsha, Y. Kravchenko, *Experimental Approach Towards Long-Range Interactions from 1.6 to 13798 km Distances in Bio-Hybrid Systems*, NeuroQuantology, 3:456-476, 2016.

Korotkov's studies and GDV device:

I.R. Bell, D.A. Lewis, A.J. Brooks, S.E. Lewis, G.E. Schwartz, *Gas Discharge Visualization Evaluation of Ultramolecular Doses of Homeopathic Medicines Under Blinded, Controlled Conditions,* The Journal of Alternative and Complementary Medicine, 9:1, 25-38, 2003.

K. Korotkov, E. Krizhanovsky, M. Borisova, D. Korotkin, M. Hayes, P. Matravers, K. S. Momoh, P. Peterson, K. Shiozawa and A. Vainshelboim, *Time dynamics of the gas discharge around drops of liquids*, Journal of Applied Physics 95 (7), 3334-8, 2004.

K. Korotkov ym., *Influence of drinking structured water to human psychophysiology*, Journal of Applied Biotechnology & Bioengineering, Volume 6 Issue 4, 2019.

J.W. Lee ja G. Pollack, *Impact of Wi-Fi Energy on EZ Water,* ScienceOpen Preprints, 3.8.2021.

T. Lewton, *A new place for consciousness in our understanding of the universe,* New Scientist, 30.3.2022.

G.N. Ling, *A new theoretical foundation for the polarized-oriented multilayer theory of cell water and for inanimate systems demonstrating long-range dynamic structuring of water molecules*, Physiol. Chem Pays. & Med. NMR. 35: 91-130.

F. Martelli, S. Torquato, N. Giovambattista, R. Car, *Large-Scale Structure and Hyperuniformity of Amorphous Ices,* Phys. Rev. Lett. 119, 136002, 2017, https://journals.aps.org/prl/abstract/10.1103/PhysRevLett.119.136002

R. Miller, *The Positive Effect of Prayer on Plants*, Psychic, April 1972. (kirjasta L. McTaggart, *The Intention Experiment,* Free Press, 2007).

Spindrift group studies in W. L. Rand's book *Reiki for a New Millennium*, B. Jain Publishers, Ltd., 2000.

E.B. Mpemba, D.G. Osborne, *Cool?*, Physics Education. 4 (3): 172–175, 1969.

L. Montagnier, J. Aissa, E. Del Giudice, C. Lavallee, A. Tedeschi, *DNA waves and water,*

Journal of Physics: Conference Series 2011; 306: 012007. https://arxiv.org/PS_cache/arxiv/pdf/1012/1012.5166v1.pdf

L. Montagnier, J. Aissa, S. Ferris, L. Montagnier, C. Lavallee, *Electromagnetic signals are produced by aqueous nanostructures derived from bacterial DNA sequences,* Interdisciplinary Sciences: Computational Life Sciences, 2009; 1: 81-90. 1

T. Morawietz, A. Singraber, C. Dellago, J. Behlera, *How van der Waals interactions determine the unique properties of water,* Proc Natl Acad Sci U S A, Jul 26; 113(30): 8368–8373, 2016, https://www.ncbi.nlm.nih.gov/pmc/articles/PMC4968748/

R.D. Nelson, *Wishing for Good Weather: A Natural Experiment in Group Consciousness,* Journal of Scientific Exploration, Vol. 11, No. 1, pp. 47-58, 1997.

Y. Scolnik ym., *Subtle differences in structure transitions between poly-L- and poly-D- amino acids of equal length in water,* Physical Chemistry Chemical Physics 8(3):333-9, 2006.

L.G.M. Pettersson, R.H. Henchman, A. Nilsson, *Introduction: Water the most anomalous*
liquid, Chem. Rev., 116, 7459-7461, 2016.

F. E. Place quotation from article G.A. Johnson, American Journal of Public Health, Vol. 1 (8), 1911.

L.N. Pyatnitskanyd, V.A. Fonkin, *Human Consciousness Influence on Water Structure,* Journal of Scientific Exploration, Vol. 9, No. 1, pp. 89- 105, 1995.

Dean Radin's and Masaru Emoto's studies:

D.I. Radin, G. Hayssen, M. Emoto, T. Kizu, Explore, September/October 2006, Vol. 2, No. 5., 2006.

D. Radin, N. Lund, M. Emoto, T. Kizu, *Effects of distant intention on water crystal formation: A triple-blind replication,* Journal of Scientific Exploration, 22(4), 481-493, 2008.

D. Radin, G. Yount, A. Delorme, L. Carpenter, H. Wahbeh, *Spectroscopic analysis of water treated by and in proximity to Energy Medicine practitioners: An exploratory study*, EXPLORE, Volume 17, Issue 1, January–February 2021, Pages 27-31. https://www.sciencedirect.com/science/article/abs/pii/S1550830720303530?via=ihub

J. Reimers, *Scorn over claim of teleported DNA,* New Scientist, 12.1.2011.

D. Riley, M. Fischer, B. Singh, M. Haidvogl, M. Heger, *Homeopathy and conventional medicine: an outcomes study comparing effectiveness in a primary care setting,* J Altern Complement Med., 2001 Apr;7(2):149-59, 2001. https://www.ncbi.nlm.nih.gov/pubmed/11327521

S.M. Roney-Dougal, J. Solfvin, *Field study of enhancement effect on lettuce seeds: their germination rate, growth and health,* J. Soc. Psych. Res. 66:129–143, 2002.

R. Roy, W.A. Tiller, I. Bell, M.R. Hoover, *The Structure of Liquid Water; Novel Insights from Materials Research; Potential Relevance to Homeopathy,* Mat. Res. Innovate. 9, 98-103, 2005.

J. Russo, K. Akahane, H. Tanaka, *Water-like anomalies as a function of tetrahedrality, PNAS, Vol. 115, No. 15, 2018.*

J. Russo, H. Tanaka, *Understanding water's anomalies with locally favoured structures*, Nat. Commun., 5, 3556, 2014.

W.C. Röntgen, *Concerning the structure of liquid water*, Ann. Phys., 45, 91-97, 1892, https://onlinelibrary.wiley.com/doi/10.1002/andp.18922810108.

M. Sammer ym., *Strong Gradients in Weak Magnetic Fields Induce DOLLOP Formation in Tap Water,* Water, 8, 79, 2016.

M. Saranpää, *Thales kaivossa,* niin & näin magazine, 1/94, 33-35.

C. Saxena, *The Concept of Water in Rig Veda* (Salilam apraketam; Rig Veda X.29.3), International Journal of Social Science & Interdisciplinary Research, Vol.1 Issue 8, 2012.

S.A. Schwartz, R.J. De Mattei, E.G. Brame, S.J.P. Spottiswoode, *Infrared spectra alteration in water proximate to the palms of therapeutic practitioners,* Explore, 11 (2):143–155, 2015.

J.M. Schurr, *Phenomena Associated with Gel–Water Interfaces. Analyses and Alternatives to the Long-Range Ordered Water Hypothesis.* J. Phys. Chem. B, 117, 7653–7674, 2013.

A. Sharma ym. *Effect of Health-Promoting Agents on Exclusion-Zone Size,* Dose-Response: An International Journal, Dose-Response: An International Journal July-September 2018: 1-8.

Y.J. Shiah, D. Radin, *Metaphysics of the tea ceremony: a randomized trial investigating the roles of intention and belief on mood while drinking tea,* Explore, 9 (6):355–360, 2013.

Y.J. Shiah, L. Shan, D.I. Radin, G.T. Huang, *Effects of Intentionally Treated Water on the Growth of Mesenchymal Stem Cells: An Exploratory Study,* EXPLORE, S1550-8307(21)00242-1, 2021. https://doi.org/10.1016/j.explore.2021.11.007

H. Tanaka, *Simple physical model of liquid water,* J. Chem. Phys., 112, 799-809, 2000.

Y. Tao, W. Zou, J. Jia, W. Li, D. Cremer, "*Different Ways of Hydrogen Bonding in Water - Why Does Warm Water Freeze Faster than Cold Water?*", Journal of Chemical Theory and Computation, 13 (1): 55–76, 2017.

C. Vega, J.L.F. Abascala, *Simulating water with rigid non-polarizable models: a general perspective,* Physical Chemistry Chemical Physics, Issue 44, 2011, https://pubs.rsc.org/en/content/articlelanding/2011/cp/c1cp22168j

P. A. Whalen, D. Kroll, P.J. Gallant, *The need for speed – rapid evolution of microbiological testing in drinking water*, October 2018, in *Microbiological Sensors for the Drinking Water Industry*

A. Widom, Y. Srivastava, V. Valenzi, *The biophysical basis of Benveniste experiments: Entropy, structure, and information in water*, Quantum Chemistry, Vol. 110, Issue 1, 252-256, 2010.

P. Wiggins, *Life Depends upon Two Kinds of Water,* PLoS ONE 3(1): e1406, 2008.

G. Yount, A. Delorme, D. Radin, H. Wahbeh, *Energy Medicine treatments for hand and wrist pain: A pilot study,* EXPLORE, 17 (1), November 2020.

J.-M. Zheng, W.-C. Chin, E. Khijniak, E. Khijniak Jr, G.H. Pollack, *Surfaces and interfacial water: evidence that hydrophilic surfaces have long-range impact,* Adv Colloid Interface Sci.,Nov 23;127(1):19-27, 2006.

www pages:

J. Blomqvist, thesis: http://urn.fi/URN:ISBN:951-45-9929-2

D. Cannon, www.dolorescannon.com

Pyramids of Egypt: https://beyondenigma.com/giza-pyramids-hydraulic-power-plants/

Madeline Ennis in Guardian: https://www.theguardian.com/science/blog/2012/apr/03/homeopathy-why-i-changed-my-mind

Grander water:

https://www.grander.com/intl-en/international/grander-wasser/wasserforschung/forschungskonzept/externe-forschung/ergebnisse-stand-2019

Homeopathy: https://magicpillsmovie.com

Jellyfish:

Scary, squishy, brainless, beautiful: Inside the world of jellyfish, National Geographic Magazine, October 2018, https://www.nationalgeographic.com/magazine/article/jellyfish-species-reproduction-feeding-ocean

How Jellyfish Control Their Lives, https://www.nature.com/articles/d41586-021-03510-6

Luc Montagnier:

Luc Montagnier, Nobel Prize Winner, Takes Homeopathy Seriously by Dana Ullman, HuffPost, 2011 https://www.huffpost.com/entry/luc-montagnier-homeopathy-taken-seriously_b_814619

Luc Montagnier 1932-2022, https://www.nature.com/articles/d41586-022-00653-y

Prahlad Jani: https://gulfnews.com/world/asia/india/india-yogi-who-claimed-to-live-without-food-or-water-dies-aged-90-1.1590498377913

G. Rein, *Homeopathy, Water and QED*, https://www.youtube.com/watch?v=NC2d6a1qFOA

Ultra magazine: https://www.ultra-lehti.com/

T. Västilä, Lääkejäämiä päätyy ympäristöön, Farmasia magazine, 10.9.2019 https://farmasialehti.fi/laakejaamia-paatyy-ymparistoon/

C. Zandonella, *Studies of 'amorphous ice' reveal hidden order in glass*, https://www.princeton.edu/news/2017/10/02/studies-amorphous-ice-reveal-hidden-order-glass

Water articles, pages etc.

Source Database: FindaSpring.com

Water conference: https://waterconf.org

WATER Journal: https://waterjournal.org

Article: Water was plentiful in the early universe https://phys.org/news/2015-05-plentiful-early-universe.html

U.S. Geological Survey: "How Much Water is There on Earth?" https://www.usgs.gov/special-topics/water-science-school/science/how-much-water-there-earth

www pages of Gerald Pollack's laboratory: https://www.pollacklab.org

Movie: Water - The Great Mystery by Hopscotch Films (Alois Gruber's interview)

Spring of Lourdes: https://www.lourdes-france.com/en/

Lynne McTaggart's www pages: https://lynnemctaggart.com

INDEX

Made in the USA
Monee, IL
04 November 2026

67384035R00128